职场翻译教材系列
Textbook Series of Professional Translation

总主编 岳 峰

科技翻译教程

岳 峰 曾水波 / 著

A Textbook for Translating
Texts of Science &
TECHNOLOGY

北京大学出版社
PEKING UNIVERSITY PRESS

图书在版编目 (CIP) 数据

科技翻译教程 / 岳峰，曾水波著 . —北京：北京大学出版社，2022.3
职场翻译教材系列
ISBN 978-7-301-31292-6

Ⅰ.①科⋯　Ⅱ.①岳⋯②曾⋯　Ⅲ.①科学技术 – 英语 – 翻译 – 教材　Ⅳ.①G301

中国版本图书馆 CIP 数据核字 (2020) 第 041547 号

书　　　名	科技翻译教程 KEJI FANYI JIAOCHENG
著作责任者	岳　峰　曾水波　著
责任编辑	刘文静
标准书号	ISBN 978-7-301-31292-6
出版发行	北京大学出版社
地　　　址	北京市海淀区成府路 205 号　100871
网　　　址	http://www.pup.cn　新浪微博：@北京大学出版社
电子信箱	liuwenjing008@163.com
电　　　话	邮购部 010-62752015　发行部 010-62750672 编辑部 010-62754382
印　刷　者	天津中印联印务有限公司
经　销　者	新华书店
	720 毫米 ×1020 毫米　16 开本　22.25 印张　518 千字 2022 年 3 月第 1 版　2022 年 3 月第 1 次印刷
定　　　价	78.00 元

未经许可，不得以任何方式复制或抄袭本书之部分或全部内容。
版权所有，侵权必究
举报电话：010-62752024　电子信箱：fd@pup.pku.edu.cn
图书如有印装质量问题，请与出版部联系，电话：010-62756370

总　序

随着翻译逐渐成为热门，翻译教材数量不断增加。传统经院教材不再居垄断地位，新教材后来居上，呈现出多样化、专业化、实用化、系列化等特点。基于语言学的传统框架逐渐被翻译人所遗忘，因为没有人会在翻译的时候想着把形容词转化成名词，或把名词转化成形容词。译者都知道，真正翻译的时候靠的是直觉，教学要意识到这点。真正实用的教学应该是以市场为导向的，一股职场翻译教程的暗流在翻译市场逐渐壮大，随着"一带一路"倡议紧锣密鼓地实施将成为主流。职场翻译教程有三个特点：

第一，职场翻译教程是以市场为导向的。翻译教材的教学内容从单一的文学翻译发展到商务笔译、旅游笔译、会展翻译、外宣翻译、应用文与公示语翻译等文宣与通稿类实践性翻译，这是一个进步。但据统计，在当今的翻译市场上文宣类翻译需求占据不到一成，而对外工程则超过三成，财经与法律占比接近三成。以经济、法律、金融、文宣、工程以及机械等各种领域为教学重点是职场翻译教程的一个重要方向。

第二，职场翻译教程开展翻译技术教学，培养学生的职业素养。在翻译职场，"十年磨一剑"也许已经不再是合时宜的话语，因为客户上门往往要求翻译几万字的内容，而且第二天就要译文。这样的翻译，单打独斗是无法完成的，需要一个团队，在一个计算机辅助翻译软件平台上，经由翻译项目管理才能如期完成。没有这方面能力的学生很难有机会进入高端机构任职。

以上两方面的内容，毋庸讳言，是高校翻译教师的薄弱之处。

实事求是地说，多数高校教师更擅长文学与文宣的翻译，支持但不熟悉翻译技术，因此职场翻译教程应该由高校教师与企业一线翻译合作编写，这也是本套教材编写的一个重要原则。

第三，职场翻译教程具有很强的实践性，因此必须注重案例在教学中的使用，通过大量模拟和实际项目来操作演练。

北京大学出版社应时代之需，推出"职场翻译教材系列"。这个系列中有全国领先的《翻译技术实践教程》，涵盖本地化翻译、译后编辑、技术传播与语料库词表，是迄今国内关于翻译技术内容较为全面的教程，由可代表国内较高水平的团队编写。也有深入浅出、文科生可以完全看懂的《翻译项目管理》。有依托新技术制作的新形态教程——一本口袋书，通过扫描一系列的二维码通往一个职场翻译教学资源库与几百分钟的教学视频。还有科技领域的一线翻译与高校教师联合编写的《科技翻译教程》，阐述科技翻译的术语、规范与易错问题，涉及数、理、化、医、电、工程、机械、采矿、制造、管理与通用类等等。该系列还包括颇为细化的《文博翻译》与《工程翻译教程》以及依托医科大学的《医学翻译教程》，涵盖中西医。"职场翻译教材系列"由校企合作、市场引领、聚焦案例，兼顾技能与素养、兼具重点与放射性，展示新技术、新内容。

岳　峰

2022年1月1日

前　　言

　　高校翻译教师译得比较多的是文宣题材的篇章，而科技翻译做得多的是语言服务产业的一线翻译，所以科技翻译教程在高校一直是比较稀缺的，科技翻译教学特别需要校企合作。鉴于此，我们以校企合作的方式进行写作，同时解决了语料、素材、案例与理论的问题。

　　科技翻译是一个很大的翻译门类，涉及的专业领域很多、很杂。所以，科技翻译跟由中高校教师讲授的占主导地位的文学、文宣等社科类翻译差别很大。要做好科技翻译，不仅需要有较好的中英双语基本功，还要有宽广的知识面，尤其是科技基础知识。但是，中国目前的文理分科模式，使得翻译行业主力军——外语翻译专业等社科专业出身的译者的科技基础知识普遍欠缺，所以在翻译科技文本时，经常会出现错误。

　　本书的写作团队就一直在做审校工作，发现国内译员在翻译专业性的科技文本时，经常感觉很吃力，译文中出现的共性问题非常多。比如，从词汇的层面上讲，很多译员喜欢把"进行"一律译成 carry out，却不知道要省译这些虚义动词，找出句子的真正动词当谓语。又如，有些英语缩写词意思难以确定，译员经常会照抄不译，不知如何查询。从句子层面上讲，很多译员喜欢机械直译，不懂得词句的转换。从对原文的理解上看，很多译员迷信原文，原文出错时没有纠正的意识，译文也跟着出错。

　　从审校时发现的典型案例入手是教学中切实可行的有效方式。我们采集日常发现的典型案例，分门别类、分析总结。集腋成裘，十几年下来积累了五十万字以上的素材。在长期的职业生涯中,将积累的案例用来培训企业译员,他们对此印象深刻,相信高校学生也会受益。

  本书在写法上与传统的科技教程的不同如下:讲述方法,注重翻译工具与辅助知识,突出难点,强调细节,规范程序,详述质检;散发着扑面而来的职场气息,深入浅出,避免经院做派,分享实战经验。本书为教育部产学合作协同育人项目"线上线下翻译技术新文科教学模式探索"结项成果。

岳　峰　曾水波

2022 年 1 月

目 录

第一章 科技笔译入门 / 1

第一节 译前准备 / 1
一、查词工具的选用 / 1
二、词典与网络解释的可信度 / 2
三、学会英英查词 / 3
四、标点符号的写法与用法 / 4

第二节 科技笔译常规 / 7
一、术语统一 / 7
二、简洁翻译 / 8
三、原文异常处的处理 / 9
四、疑难点与不确定译法的处理 / 9
五、需使用固定译法的内容 / 9
六、大小写、数字、项目编号与文件编号等的翻译 / 10
七、译后自检 / 13

第二章 应用类文件翻译要求 / 15

第一节 简洁化翻译的原则与方法 / 15
一、省略次要成分 / 17
二、使用缩写词 / 18
三、习惯性省译 / 19
四、其他省译情况 / 21
五、使用后置或前置定语 / 22

六、用单词替代从句或短语 / 23

七、词性转换 / 25

八、省略共同成分 / 28

九、说法变换或句式转换 / 28

十、不啰唆、不绕弯翻译 / 30

十一、适当多用短句，少用长句和复杂句 / 32

十二、图片、表格内容的简化译法 / 33

第二节 时态、语态和句式 / 36

一、中译英 / 36

二、英译中 / 43

第三节 行文表达 / 49

一、正式用语（避免口语化）/ 49

二、位置与语序 / 50

三、目标语行文习惯 / 55

第三章 中文理解与表达 / 64

第一节 翻译相关中文基础知识 / 64

一、中文连词搭配 / 64

二、易用错的中文词 / 66

三、易误译的中文词 / 70

第二节 常见中国特色表达 / 80

一、中国特色动词 / 80

二、中国特色名词、形容词及虚词 / 92

三、其他中国特色表达 / 98

四、其他中国特色疑难词 / 107

第四章 英文理解与表达 / 117

第一节 英文遣词造句注意点 / 117

一、英文小词的用法 / 117

二、独立主格结构 / 134

第二节 词义辨析 / 138

 一、一词多义 / 138

 二、近义辨析 / 146

 三、"意外"词义 / 156

 四、英文常见符号和包含数字的词 / 160

第五章 疑难点解析 / 161

第一节 英文缩略词的缩略规则 / 162

 一、一般缩略规则 / 162

 二、其他缩略规则 / 167

 三、缩略词的大小写 / 171

 四、名词缩写的复数形式 / 172

第二节 查词法 / 173

 一、英文查词 / 173

 二、中文查词 / 181

 三、结合逻辑与知识面 / 188

第三节 句子理解 / 193

 一、关键词理解 / 193

 二、原文理解陷阱 / 198

 三、习惯性不规范表达 / 210

 四、填空式句子 / 211

 五、复杂句式理解 / 215

 六、逻辑及背景知识 / 232

第四节 回译（返译）/ 242

 一、讲话的回译 / 242

 二、文件名的回译 / 245

第五节 原文纠错内容 / 248

 一、英文原文纠错 / 248

 二、中文原文纠错 / 253

第六节　专有名词的译法 / 255
　　一、专有名词译法的确定 / 255
　　二、汉字文化圈文字的相通性 / 258

第七节　其他 / 262
　　一、非专业用词的翻译 / 262
　　二、非正常词义 / 265
　　三、原文同位语或解释性表达的翻译 / 266
　　四、英文名词的单复数 / 267

第六章　其他相关技巧与知识 / 269

第一节　各种文字的辨识 / 269
　　一、繁体字与异体字 / 269
　　二、一字读两音 / 270
　　三、手写字 / 270
　　四、艺术字 / 270
　　五、不完整（截断或模糊）文字的猜测 / 271

第二节　数字、量级与单位表达法 / 272
　　一、阿拉伯数字的表达法、数字量级 / 272
　　二、货币单位符号和货币量级 / 273
　　三、单位的译法 / 274
　　四、常见标准、指令、条例、协会的缩写 / 281

第七章　各行业语篇翻译分析 / 284

第一节　通用类 / 284
　　一、中译英 / 284
　　二、英译中 / 288

第二节　管理体系 / 293
　　一、中译英 / 293
　　二、英译中 / 302

第三节　机械工程制造 / 307
　　一、中译英 / 307

二、英译中 / 312
第四节　物理、电子电工 / 316
　　一、中译英 / 316
　　二、英译中 / 321
第五节　化学化工、生物、医学医药 / 327
　　一、中译英 / 327
　　二、英译中 / 330
第六节　石油地质采矿 / 334
　　一、中译英 / 334
　　二、英译中 / 337

第一章　科技笔译入门

科技笔译要求高，行规多，注意点密集，因此学习者不要急于求成，应打好基本功，很多细则需要从头学习。

第一节　译前准备

一、查词工具的选用

必须学会借助网络词典和网络搜索引擎。有经验的译员常同时使用多个网络词典与搜索引擎。下表为常用的类型，供参考：

表 1.1　笔译常用网络词典和搜索引擎[①]

词典名	网址	简介
金山词霸	http://www.iciba.com/	中文互联网多语词典，支持中、英、日、德、法五种语言查询，约含 200 本词典，覆盖几十个专业领域。并且包含中文、英文真人发音和常用单词的视频讲解。
海词	http://dict.cn/	拥有全面的词库。智能词库生成技术保证海词收词的权威性和全面性。目前拥有几百万不重复词条，包含各行各业的术语和生僻词汇。在不重复词条数目上，是同行网站拥有的传统词库的几十倍。
有道	http://dict.youdao.com/	数千万条例句可一键查询，还可根据单词释义选择相关例句，帮您更加准确地理解单词语境，活学活用。

---

① 网址检索时间为 2020 年 9 月 1 日。

（续表）

词典名	网址	简介
必应词典	http://cn.bing.com/dict/	微软中、英文智能词典。不仅可提供中、英文单词和短语查询，还拥有词条对比等众多特色功能。
通译典	http://tdict.com/	容量大、功能全，有海量的中、英翻译语料库，有不重复词条近500万条，容量居所有词典之首，并与大多数网络上可用词典资源链接。
灵格斯		本工具可自主添加词典，在网络断线的情况下，也可凭借所添加的词典查词。
韦氏在线词典（纯英文）	https://www.merriam-webster.com/	纯英文词典，韦氏词典在词义定义的准确性方面享有盛誉，在发音、帮助、近义词、用法和语法技巧方面颇为可靠。
百度搜索引擎	https://www.baidu.com/	可用于搜索中文词条，也可用于搜索较常见、较简单的中英文词句互译。
Bing 搜索引擎	http://cn.bing.com/	分国内版和国际版。可搜索到与翻译相关的大部分疑难点。

二、词典与网络解释的可信度

中国有句古话"尽信书不如无书"。各类型资讯，纸质或网络的，可信度不同。我们根据经验评价如下，供读者参考：

表1.2　词典与网络的可信度

信息来源	错误率	举例	原因
权威词典	错误率极低，文字错误率估计在百万分之一以下，甚至几乎无误。比如《辞海》、孙复初主编的《新英汉科学技术词典》可信度极高。	《辞海》《新英汉科学技术词典》《牛津词典》《朗文词典》《韦氏词典》《新英汉词典》《汉英大词典》	经反复审核、多次再版
专业词典	错误率很低，但有些专业词典出版时间不长，会有极少量的错误，只有高水平专业人士才能发现。	各行业的专业词典	经多次把关反复审核，有些已多次再版

(续表)

信息来源	错误率	举例	原因
教科书与专业书籍	发行量大、口碑好的教科书与专业书籍错误率极低,文字错误率估计在百万分之一以下,只有高水平专业人士才能发现	国内多数基础学科的畅销版本的教科书,如国内知名出版社义务教育阶段的教科书、本科或专科基础学科的教科书	反复审核,有些已多次再版
	专业性强、发行量较小的教科书与专业书籍准确率会稍低些,甚至普通读者也能发现文字错误,但内容错误极少。	高校专业性较强、发行量较小的教科书	出版前反复审核,通常再版次数较少
个人专著	文字错误率估计为万分之一左右,但内容准确率参差不齐。		出版前反复审核
网络词典	多数网络词典准确率不够高,估计词条解释的错误或不当率在千分之一左右,文字错误率更高。	金山词霸、海词、有道、必应词典	审核不严,有些是网友加上去的解释
网络	网络搜索结果准确率视来源而定,整体错误率偏高,经常在百分之一以上。有些网络问题的回答很不可靠。网络论文也有不少低质量的。专题文章则准确率较高。	百度主要用于查中文概念或简单中英互译。	随便什么人都可回答问题或发表文章。

译者应该基于多种资讯,做出自己的判断。

三、学会英英查词

英汉词典多数只提供所查词条的中文表述,英英词典会提供与该词条相关的具体信息,这可以帮助译者判断。就像中文近义词之间的词义必须通过查中文词典加以区别一样,在选择英文用词,或者理解英文词的准确词义时,通常也不能靠中文解释来加以辨别,而是要查纯英文词典,或者到网络上用英英查词法了解其区别。比如，要了解dedicated和specialized的区别，可以在Bing搜索引擎上查"dedicated specialized difference"，从而发现其细节差别。另外，通过在网络词典里查数个近义词的解释，也可以了解到近义词之间的区别。

四、标点符号的写法与用法

英文标点符号与中文标点符号有较多的不同点。很多新译员在初次翻译时可能会混用。值得注意的是，在英文里使用中文标点符号，除了不符合规范外，还会导致国外电脑无法显示，在中文标点符号出现处显示空格或乱码。

1. 英文里不得使用的全角中文标点符号：

，　；　。　·　、　！　——　——　《》　〈〉　（）　" "
' '　？　：　……

中文标点符号是全角的，而英文标点符号是半角的，格式不一致。

以上标点有些是中文所特有的，如顿号、书名号、间隔号（即用于日期的月日之间或人名中间的黑点）和着重号（汉字下方的黑点），不能在英文中出现。

2. 中文标点符号用法

中文里使用英文标点符号，虽然不会导致生成乱码，但也是不规范的。而且，英文标点符号只有半个字符的空间，会显得排版很局促。如果文本中中英文标点混用情况严重，可以用电脑的替换功能（F7）进行修改。

另外，很多国人常犯的中文标点符号使用方面的错误，一是顿号的使用，二是逗号的使用。中文并列小成分之间一般都用顿号隔开，而不是用逗号，比如"业主均为水暖、电力、网络等提供和安装必要管道、配件和阀门等"。而逗号则不仅可用于分句之间的停顿，也可用于主语与谓语之间、句子内部动词与宾语之间，以及其他必要的语气停顿（如状语后的停顿）处，如"进出配电箱/盘的端子，必须加强绝缘并采取固定措施"。

3. 英文标点符号写法规则

英文标点主要包括：

, 　.　 ; 　: 　- 　' 　'' 　' 　' 　" 　" 　? 　! 　~ 　( ) 　[ ]
{ }

几个易错的英文标点符号的写法：

（1）前引号（包括单引号和双引号）和前括号前面空一格。如：The United

States of America is abbreviated as "USA", 又如: MBA (Master of Business Administration) was very popular at that time.

但前引号和前括号出现在段首时, 前面不空格。

（2）连字符前后都不空格, 如: X-ray。

（3）单词中的分隔符前后都不空格, 如it's a good idea。

（4）数字里的小数点和千分撇前后不空格, 如1,234.56。

（5）后引号后面如果有逗号、句号等其他标点符号, 则后引号后不空格。

（6）缩写词后面的一点在句中时, 后面空一格, 在句末时不空格。

4. 符号翻译注意点

原文段落最后一句或者独立成段的句子的句末缺少标点符号, 或者使用了不正确的标点符号时, 译员必须补充或修正。原文明显用错的, 翻译时应改正过来。译文不必完全延用原文的标点符号, 可以根据译文断句的需要自行决定。但是也不可过分修改。

5. 特殊符号的翻译

英文里有时会出现一些特殊符号, 有些是原文里有特定意义的, 有些则是源文件经格式转化后出现的错误写法。后者一般会让人感觉异常或无法理解, 这时需查看原文, 采用正确原文译出。英文里有特殊意义的常见符号意思如下:

表1.3 特殊符号的意义

符号	意思	举例	说明
&	= and	Operation & Maintenance（运行与维护） R&D 研发	& 一般当作 and 译出, 不宜在中文语境里出现
@	= at, against	（1）12345678@qq.com （2）=at 或 against。如: resistance @20℃（20℃时电阻）	（1）不译 （2）@ 做 at 或 against 使用时, 一般要译出
1.234.567,89	= 1,234,567.89	前者是德、法等欧洲国家的写法, 和中国用法刚好相反。	请了解各种数字写法

（续表）

符号	意思	举例	说明
X 或 x	（1）X = cross、extra、trans、Christ 等 （2）X 代表倍数 （3）X 代表乘号	X-conn = cross connection（交叉连接） XL = extra large（特大号） Xmas = Christmas（圣诞节） a 10X magnifier（10倍放大镜） Standard container: 5. 86X2.13X2.18（标准箱：5.86×2.13×2.18）	X 或 x 的这些用法视具体情况决定译与不译 X 代表乘号时可用 * 代替
/	/	3/8" = $\frac{3}{8}$"（八分之三英寸） A/C = alternating current（交流电） L/C = letter of credit（信用证）	/ 表示除号、分数等，意思不明时可查网络确认
V 和 X	V = √ X = ×	常出现在表单证件类以及其他相关文件中，用于选项的选择。 （1）Authorized representative: Yes V No .（是否授权代表：是 V 否） （2）Use of intellectual property X .（有无使用知识产权 X .）	V 和 X 的这些用法可照抄，也可抄成√和 ×
Ø、φ	直径或角度（相位）符号	（1）Ø15 pipes（直径 15mm 的管道） （2）3 Ø = 3 phases（三相） （3）Ø60º（60º 角）	大写为 Ø，小写为 φ
2	= two 或 to	Y2K = Year 2000（二〇〇〇年） B2B = business to business（企业对企业；商家对商家）	
m2 m3 km2	= m^2 = m^3 = km^2	The area of this square is 4m2.（这个正方形的面积为 $4m^2$）	这里的数字，翻译时必须转化成上标
4.6X106 7.8X10−3 之类表达	X 为乘号 6 和 −3 为指数	= 4.6×10^6 = 7.8×10^{-3}	数字乘方中的指数必须写成上标

（续表）

符号	意思	举例	说明
°　′和″		（1）The angle between earth equator and the ecliptic is 23°26′21″.（地球赤道与黄道面夹角为23°26′21″。） （2）Height of Michael Jordan: 6′6″.（迈克尔·乔丹身高6英尺6英寸。） （3）The time interval is 10′30″.（时间间隔是10分30秒。）	°　′和″表示角度时分别指度、分和秒 ′和″表示尺寸时分别指英尺和英寸 ′和″表示时间时分别指分和秒 <u>但在实际翻译时，这些符号都可以不译</u>
~	~	（1）~90 degree（约90度） （2）220V~50HZ（交流电压220V，频率50Hz）	在数字前理解为"约" 在电压后表示"交流" 直流则表示成一横杠，如：110V-（直流电压110V）

本节课后练习

请回答下列问题：

1. 笔译常用的网络词典和搜索引擎有哪些，请说明其各自特色。
2. 网络词典、纸质词典、教科书、个人专著与网络搜索的准确率如何？
3. 你见过什么样的中英文标点混用？

第二节　科技笔译常规

本节是对译员的一般性的要求。

一、术语统一

术语指文件中出现的与各专业相关的专业词汇，或者为严谨起见而必须加以统一的非专业词汇，其中包括但不限于文件中出现的需统一的文件名、人名、地名、职称名、组织机构名等。术语统一是多人合译同一文件过程中非常重要的要求。如果各译员术语不统一，后面无论是分别修改或统一修改，都会浪费很多时间，有时甚至无法统一。

因此，在此提醒大家养成良好习惯，在首次碰到可能反复出现的词语（包括专业或非专业术语、人名、地名、组织机构名、文件名等）时就做好统一。养成习惯，节省时间。具体要求如下：

如已提供术语或参考文件，则严格按术语表（非双语对照的术语表必须译出）或参考文件的译法翻译。译员如有不同意见，可以做批注说明后采用自己的译法，或者跟项目派单人员沟通后确定译法。如未提供术语或参考文件，则同一译员对同一意思的同一术语前后译法应保持一致。

第一次碰到可能需要统一的词时，查看一下其他人的译法，如果其他人已有唯一合理译法，就采用其译法（用MMQ时可用Ctrl+K键查看，或者用MMQ的筛选功能查看，用其他工具时也可使用相应办法查看）。如果有数个合理译法，或者已有译法都不合理，可讨论确定统一译法。如果还没人译过，就自己确定一个译法并保持自己的术语统一。

在参与大项目的合作翻译时，对于内部译员，应及时关注大家在项目讨论组里的发言，以及时做好术语统一或遵守其他相关规定。对于外部译员，应将需要统一的术语及其译法以表格形式列出，以便审校修改不统一的术语。

另外，在多人合译时，为提高整体质量，也为提升译员水平，一般都要在QQ等聊天工具里建立讨论组，讨论相应问题。译员碰到疑难点时，可以在讨论组里提问。但是，为提高解决问题的效率，请尽量避免以贴图的形式提问，而是把相关文字复制粘贴出来，同时说明是在哪个订单哪个文件里出现的问题。否则，会导致答疑者在必要时，无法直接复制文字，还要另外打字，或者找不到相应文件了解上下文，从而浪费时间。而且，这样的提问在历史记录里还没法查找。

二、简洁翻译

实用翻译都对译文质量有较高要求，好译文通常更简洁，字数更少，而我们与译员结算时，却经常按中文字数计算。因此，为保证较高的翻译质量，防止低质量的啰唆译文反而获得更高报酬，需郑重要求，除个别很口语化的文件外，不管是中译英还是英译中，都尽可能简洁用词，多用正确恰当的书面语或正式用词，少用口语或非正式用词。如果译文太过啰唆，有时会被要求返稿修改。

三、原文异常处的处理

当原文是 PDF 或图片转化过来的 WORD 文档时，或在 Trados、雪人或 MMQ 等翻译辅助软件里碰到感觉异常的词句时，记得查找一下原文，看是不是转化出错，或者转化后排版出错（如一词或一句被分成两行以上，标点错漏，或者不该合并的词句被合并）。

由于使用软件（如 MMQ）做翻译时，会因软件功能不完善而出现一些译文与原文对不上号的现象，如多译、少译、漏译、不对应译文、原文一句被分为多句、原文多句被合并为一句、原文或译文所处位置不对、译文位置无法调整等现象。所以，要求各译员在译完自己的文件后，快速对着原文浏览一遍自己的译文，碰到异常的地方，认真检查一下，有错误或不当之处，及时改正。

另外，翻译过程中碰到不确定的内容，或者感觉原文明显有问题的地方，请及时与派单人员沟通，或者把原文和译文分别标黄色底色或做批注，以便审校人员解决。

四、疑难点与不确定译法的处理

对于在翻译过程中碰到的疑难点或不确定译法，译员如果无法自己解决，可将其加标底色或做批注，以便审校人员及时解决（加标底色更好，建议标成黄色，不建议用深色，以免看不清文字）。

五、需使用固定译法的内容

有些内容不能自己译，而是要在网络或者纸质书籍上找现成译法。这些内容包括但不限于如下：

1. 知名人士姓名

翻译时须译成相应的中文名或英文名。对于无英文名的华人，采用拼音翻译时，如果不确定，就要查网络或纸质词典确定拼音。对于无中文名的外国人，译成中文时，一般则采用规范的音译法。

2. 公司、企业等组织机构名称

这些单位一般都有既定的英文，一般要上网查现成译法，无现成译法时才自己译。

3. 职务、职位、职称

国内外没有一套完全合理的标准，只有一些国际性大公司有相对成熟合理的职务职称说法。对于没有标准译法的职务职称，可以根据上下文翻译，同时结合合译者的译法，予以统一。

4. 书籍、文章名

通常也要上网查找既定译法。如果没有统一的既定译法，可以选用网络上相对合理的译法，或者自己以合理的方法译出。

5. 行业标准或国家标准规范名称

不论是中文标准规范，还是英文或其他语种标准规范，只要是比较权威比较常见的，通常在网络上都能找到统一的既定译法，翻译这些标准或规范时，必须使用这种既定译法。如果没有统一的既定译法，可以选用网络上相对合理的译法，或者自己以合理的方法译出。

6. 直接引用的内容

直接引用的书籍、文章、标准、规范等的内容或人物讲话等（通常加引号），在译成外文后，如果要再回译成源语言（比如，某中国名人的讲话译成英文后再回译成中文），需查找其原始表达，而不是自己翻译。

以上内容之所以要找现成译法，主要是为了防止译文与原文对不上号，或者防止翻译完文件后，要再次回译（即从译入语再次译成译出语）时，无法准确还原。

六、大小写、数字、项目编号与文件编号等的翻译

1. 大小写

一般的中译英文件，标题首字母大写到节，或者大写到约定的其他级别。

此外，一些文件、书籍、报刊、规范/标准、人名等专有名词的实词也应该首字母大写。甚至包括合同、协议、章程、财务报告里的一些具有专门意义的名词，其实词首字母都应该大写，如：本合同（the Contract）、发包人（the Employer）、章程（the Articles of Association）、董事会（the Board of Directors）。

2. 数字

（i）原文里的时间、金额等数字，绝对不能抄错或译错。

（ii）阿拉伯数字中间不得使用中文逗号，而且中间不换行，以免误解。

（iii）金额和数字的表达，必须符合目标语行文习惯。

3. 项目编号

项目编号一般按以下原则翻译：（1）可以不译的项目编号原则上都不译。项目编号多数都是阿拉伯数字或英文字母，这类编号不论中译英还是英译中，一般都照抄不译。

中文	English	说明
（1）、（2）、（3）	(1)、(2)、(3)	照抄，但英文里不准出现中文字符
i、ii、iii、iv	i、ii、iii、iv	
a、b、c、d、e	a、b、c、d、e	

（2）以中文一、二、三、四……表达的项目编号，因译成one、two、three、four…太过麻烦，一般按以下方式翻译：

第1章、第2章、第3章	Chapter 1、Chapter 2、Chapter 3	不论中译英还是英译中，译文一般采用1、2、3为章的编号
第一章、第二章、第三章		
第1节、第2节、第3节	Section 1、Section 2、Section 3	不论中译英还是英译中，译文一般采用1、2、3为节的编号
第一节、第二节、第三节		
第1条、第2条、第3条	Article 1、Article 2、Article 3	不论中译英还是英译中，译文一般采用1、2、3为条的编号
第一条、第二条、第三条		
一、二、三、四……	I、II、III、IV……（罗马数字）	注意，IV 和 VI 别弄混，并记住十以上的罗马数字写法

当将一、二、三、四等中文数字译成I、II、III、IV等罗马数字时，其对应关系如下：

罗马数字	I	II	III	IV	V	VI	VII	VIII	IX	X	XI	XII	XIII	XIV	XV
阿拉伯数字	1	2	3	4	5	6	7	8	9	10	11	12	13	14	15
罗马数字	XVI	XVII	XVIII	XIX	XX	L	C	CL	CC	CCL	CCC	CD	D	M	
阿拉伯数字	16	17	18	19	20	50	100	150	200	250	300	400	500	1000	

但要注意：① 中译英时，罗马数字编号不得从WORD的"插入"命令中直接插入，否则会变成中文字符。② 中译英时，项目编号与后面的文字之间应该空一格或两格。

4. 文件编号等

文件编号或物品名称编号里的中文文字，通常不必逐字逐词翻译，而是用拼音首字母译出，也可译成拼音全拼。如：

沪司发律管（1999）49号　H.S.F.L.G.(1999) No.49

又如：

The specification IEC 60034-14　（国际电工协会）规范IEC 60034-14

之所以这样译，是因为在将此类编号里的中文逐字逐词译成英文后，通常看不出实际意思，反而增加读者的困扰。而纯粹用拼音首字母译出后，既保留了中文原文的文字信息，又让国外读者知道这只是个编号，没有特殊含意。

5. 英文缩写的翻译

英文使用缩写通常是因为其全称太长，读写不便，不得已才用缩写。但缩写之后，往往在后文见到缩写时，又无法立即理解其意思，这就不可避免地导致阅读效率降低。由于中文的简洁性，这种用词篇幅过长的情况几乎不存在。如：NATO全称为North Atlantic Treaty Organization，中文意思"北大西洋公约组织"，简称"北约"；又如：UNESCO全称为United Nations Educational, Scientific and Cultural Organization，中文意思"联合国教育、科学及文化组织"，简称"联合国教科文组织"。

所以，以后大家做英中翻译时，英文缩写尽可能译成中文，如果全称太长，就用中文的简称。当然，鉴于中外交流的现状，也会有些例外——某些行业里被普遍接受的英文缩写词，如 DNA、CPU 之类的，有时候可以不译，译出来反而让人不熟悉。尤其是在生物、医学、电子和计算机等英文缩写较多的学科，在翻译之前，译员必须要判断是不是属于在中文里普遍使用缩写的情况（可通过上网查看确定，但必须看比较权威的资料来源），如果不是，就直接译成中文全称或中文缩写。

七、译后自检

译员必须养成译后自检的习惯。译后自检，其目的是消除英文译文里的中文字符，中文译文里不应出现的英文标点符号，拼写语法错误，数字、编号、性别、人名、地名、职务、职称、组织机构名等的错误以及因疏忽产生的其他错误，同时也是解决翻译中碰到的疑难点的一个好时机。

怎样检查英文中是否有中文字符？可在WORD"查找和替换"的"查找内容"里输入中文字符的通配符[!^1-^127]，然后选中"使用通配符"，查找后全部突出显示（建议用黄底色显示），最后逐个检查修改。

数字错误是指译文数字与原文不一致，包括数字和单位的错误。这些错误，在译员看来可能是小问题，但对客户来说，却是极严重的问题。因为数字往往是文件里的重要信息，是绝对不能出错的。要消除译文里的数字错误，一般可在"查找和替换"的"查找内容"里输入阿拉伯数字的通配符^#，查找后全部突出显示（建议用黄底色显示），最后逐个检查。这种方法对非阿拉伯数字无效。

拼写和语法错误，是指文字的拼写和行文中的语法错误（尤其是英文的拼写和语法错误）。这类错误，通常是客户比较容易发现的问题点，也是译员比较容易忽略的，只要使用WORD的拼写与语法检查功能即可整本消除。在WORD2007以上版本的"审阅"工具中，找到并点击"拼写和语法"，即可调出拼写和语法检查对话框，开始检查。也可以按快捷键F7迅速调出该对话框。常见的语法错误，除了笔误，主要有冠词、名词单复数、第三人称单数错误等。可查阅相关语法书进行修正，如各版本牛津词典或朗文词典的附录以及《张道真实用英语语法》等书。

本节课后练习

请回答下列问题：
1. 术语统一的要求是什么？

2. 怎样做到简洁翻译?
3. 原文异常处该怎么处理?
4. 对于疑难点或不确定译法的地方,该做如何处理?
5. 需使用固定译法的内容有哪些?
6. 请说明大小写、数字、项目编号与文件编号等的翻译原则。
7. 是不是所有英文缩写都要译成中文?如果不是,哪些英文缩写可以直接在中文里使用?
8. 请列举出英文里常用的一些特殊符号的意思以及德、法等欧洲国家数字写法与中国的区别。
9. 译后自检包括哪些方面?请说明各方面内容的自检技巧。

第二章　应用类文件翻译要求

应用类文件最主要目的是传达知识技术等有用信息,其最基本原则是准确、合乎目标语行文习惯。但是,由于现代社会高度追求效率,人们希望通过尽可能少的语言准确地了解到尽量多的信息,所以应用类文章也必然要求行文简洁明了,不浮夸修辞,不啰唆纠缠,能简单说清的,就不绕弯,能用一个词的,就不用两个词或更长的词组,能用短词组的,就不用长单词,能用小词的,就不用大词,并少用虚词。

总之,应用类翻译的原则就是准确、简洁、明了,并具有一定的正式性与美感。也就是说,在某种程度上,应用类翻译就是简洁化翻译。

本章内容首先介绍简洁化翻译原则和简洁化翻译方法,然后再介绍时态、语态和句式以及行文表达方面的原则和技巧。

本章重点不在理论阐述,而是想让读者尽量多地学习到翻译技巧。

第一节　简洁化翻译的原则与方法

在英文简洁性方面,做的最好的,应该是新闻界。新闻翻译,就是简洁化翻译的典型,其特点是:**具体、准确、简明、通俗、生动**[①],避免使用复杂句和长句。

应用类英文与新闻类英文虽然在形式上有较大区别,但在传递信息的具体、准确、简明方面是具有很大共性的。因此,想做好应用类文件的翻译,就必须对新闻翻译有一定的了解。请看下例:

---

① 许明武,《新闻英语与翻译》,北京:中译出版社(原中国对外翻译出版公司),2003:19。

例2.1：

原文：中国神舟十一号载人飞船与天宫二号空间站成功对接

原译：China's Shenzhou-11 Manned Spacecraft Successfully Docks at the Tiangong-2 Space Station

改译：China's Shenzhou-11 Docks at Tiangong-2 Space Station

分析：本例属于科技类文章标题。一般而言，文章标题应比正文更简洁，所以，译文<u>应尽可能简短但又尽可能蕴含更多信息</u>。本例所用到的简洁翻译方法包括：

（1）省略虚词：定冠词the被省略；

（2）省略非关键词：manned spacecraft和successfully被省略；

（3）使用短词：将"对接"译成短词dock at；

（4）数字替代单词：将"神舟十一号"译成Shenzhou-11；

与新闻类英文的行文方式类似，应用类文章常见的中译英方法还有短词替代长词、单词替代短语、词性转换、说法转换与句式转换等。这些方法将在本章第二节加以介绍。

类似的简洁化译法，还经常出现在<u>工程项目、组织机构等</u>的名称以及表格等其他需要简洁翻译的地方。如：

例2.2：

原文：

表1.1-1　××石化公司装置信息一览表

序号	车间	范围
1	蒸馏车间	350万吨南蒸馏、300万吨北蒸馏
2	催化气分车间	100万吨催化裂化、50万吨气分、气体脱硫
3	重整车间	80万吨连续重整、15万吨芳烃抽提、120万吨汽油加氢、20万吨航煤加氢、芳烃罐区
4	尿素车间	10万吨尿素脱蜡

译文：

Table1.1-1　for Information of Devices of XX Petrochemical Company

SN	Workshop	Scope
1	Distillation Workshop	3.5M t/a South Distillation, 3.0M t/a North Distillation
2	Catalytic Cracking & Gas Separation Workshop	1.0M t/a Catalytic Cracking, 500k t/a Gas Separation, Gas Desulfuration

（续表）

SN	Workshop	Scope
3	Reforming Workshop	800k t/a Continuous Catalytic Reforming, 150k t/a Aromatics Extraction, 1.2M t/a Cracked Gasoline Hydrogenation, 200k t/a Aviation Fuel Hydrogenation, Aromatic Tank Area
4	Urea Workshop	100k t/a Urea Dewaxing

分析：项目名称与文章标题一样，都要尽可能简化翻译。其简化方法除了前例介绍的之外，还包括使用：

（1）数字：多用阿拉伯数字和符号（如：k［千］、M［百万］、B［十亿］）表示。

（2）单位：多用字母或缩写词表示。如：

350万吨南蒸馏 3.5M t/a South Distillation

下面，对简洁化翻译的具体方法予以举例说明。

一、省略次要成分

次要成分，是指句子中可以省略而不影响句子整体意思的成分。这些成分的意思，通常已经隐含在上下文中，或者根本没必要译出。比如：

例2.3：

原文：当时那种金属必须在低温下熔化。

原译：It was necessary that the metal should melt at a low temperature at that time.

改译：It was necessary that the metal should melt at a low temperature.

分析：原译把"当时"译出来了，改译后则把"当时"省译了。因为it was的时态已经暗示其时间为过去。这种动词时态所隐含的时间，经常会被省略。

例2.4：

原文：左击变量文本，打开的弹出菜单显示有变量文本的对应类型选项。

原译：Left-click the variable text, the pop-up menu opened up will show it has a corresponding type option to the variable text.

改译：Left-click the variable text, the pop-up menu opened will show the type option corresponding to the variable text.

分析：本例中的"显示有"是一个整体，相当于"显示"或"会显示"，这是中文的一种特殊表达方式，可以省译"有"字。但原译却把"显示有"理解成"显示其拥有"，译成show it has。像这种中式表达，需要译员必须充分理解原文的意思，而不是凭感觉翻译。

二、使用缩写词

在中译英时，一般英文篇幅都比中文长一半左右，如碰到中文使用缩写词，或者成语、典故、古语等含义很丰富的词，英文篇幅就会更长。所以，在中译英时，有些中文词汇，尤其是约定俗成的中文词，或者在上下文背景下可判定为术语的词，如果译成英文全称太过冗长，则可采用英文缩写。

例2.5：

原文：研发部是公司的核心部门，肩负着研制、开发新产品，完善产品功能的任务。

原译：Research and Development Department is the core department of a company, shouldering the tasks of research and development of new products and improvement of production functions.

改译：R&D Department is the core department of a company, shouldering the tasks of R&D of new products and improvement of production functions.

分析："研发"是企业的一个重要功能，研发部是企业的重要部门，"研发"在英文里普遍缩写为R&D。类似地，企业的其他常见功能或功能部门也经常采用缩写。又如：

HR = Human Resourses：人力资源（人资）（部）

QC = Quality Control：品质管理（品管）（部）

QA = Quality Assurance：品质保证（品保）（部）

例2.6：

原文：所采用电缆应具有耐候性，适合热带气候条件，并具有可承受1,100V电压的聚氯乙烯绝缘层。

译文：The cables shall be weatherproof and suitable for tropical climate having PVC insulation of 1,100V grade.

分析：本例将"聚氯乙烯"直接译成PVC，即polyvinyl chloride。在英文文章里，常见的自然科学相关名词经常采用缩写，这样可以有效地缩小文章的篇幅，尤其是在大量使用自然科学名词的文章中。比如：

  HDPE = high density polyethylene：高密度聚乙烯

  PBDEs = poly brominated diphenyl ethers：多溴联苯醚

  CMOS = complementary metal oxide semiconductor：互补式金属氧化物半导体

  XOR gate = exclusive-OR gate：异或门

  DNA = deoxyribonucleic acid：脱氧核糖核酸

例2.7：

原文：兹将本报告郑重提交给中国国家海洋局①，以更新康菲石油中国有限公司与中国海洋石油总公司（PL 19-3的联合投资方）自从2011年6月溢油事件发生以来由国家海洋局所发布的各项指令有关方面所取得的成绩。

译文：This report is respectfully submitted in order to update the SOA on the achievements that COPC and CNOOC, the PL 19-3 co-venturers, have made in regards to the various directives that the SOA have issued since the onset of the oil spill events in June, 2011.

分析：本例中用到了多个缩写，且所有缩写都是组织机构名，其中：

  SOA = State Ocean Administration, People's Republic of China：中国国家海洋局

  COPC = ConocoPhillips China Inc.：康菲石油中国有限公司

  CNOOC = China National Offshore Oil Company：中国海洋石油总公司

不过，需要注意的是，这种缩写，通常是在目标读者对这些缩写都比较熟悉的情况下才使用。

三、习惯性省译

所谓的习惯性省译，是指在中文表达里出现，但在英文表达里不必译出的词。这些词主要包括：

（一）型、式、化、性、类、法、级等表示归类分级的词。这些词与中文量词类似，在英译时经常不译出，如：

---

① 机构改革前名称。

中文原文	不合理译法	简洁译法	说明
微型账户	micro type account	micro account	型：样式
新型拓扑	new type topology	novel topology	型：样式
前瞻式分析	forward analyses	prospective analyses	式：样式，样子
集约化经营	intensification operation	intensive operation	化：性质或状态的改变
恶性肿瘤	malignant property tumor（错译）	malignant tumor	性：类别，范畴
蟹类	crab class	crabs	类：种类
生物接触氧化法处理生活污水	domestic sewage treated by biological contact oxidation process	domestic sewage treated by bio-contact oxidation	法：方法
指数级增长	exponential-class growth	exponential growth	级：层次、等次

需要指出的是，正如上表最后一列对型、式、化、性、类、法、级的解释，这些词并非关键词，很多时候只是用在中文形容词后补充说明归类、方法或层次，如果在中文里省掉这些词，通常意思表达也不受影响，只是不符合中文表达习惯。其中，型、式、性、法这几个词，由于通常表示归类样式，实际上可以用"的"代替。因此，在英文中，这些词经常可以不译。或者说，这种归类、方法或层次已经包括在英文形容词中，比如在上表的micro（微型）、novel（新型、新颖）、prospective（前瞻式）、intensive（集约式）、malignant（恶性、有害）、exponential（指数级）。

但是，型、式、化、性、类、法、级等表示归类方法或层次的词，并不是一律不译，有些时候习惯性译出，而且在强调某种归类、方法或层次时也需译出。比如：II型糖尿病（type-II diabetes）、中国式人际关系（Chinese-style relationship）。

（二）"电""钢""值"等的省译

在涉及电学、钢铁制品等的行业，很多用电的产品或者钢制品，由于在上下文中默认是用电的或钢制的，如果不引起误解，"电""钢"通常不译。比如：

中文原文	不合理译法	简洁译法
电源	power supply source	power 或 power supply
电灯	electric light/ lamp	light/ lamp

（续表）

中文原文	不合理译法	简洁译法
电动机	electric motor	motor
电压	electric voltage	voltage
钢丝	steel wire	wire
钢丝绳/刷	steel wire rope / brush	wire rope / brush
钢管	steel pipe	pipe
钢板	steel plate	plate

另外，在表示某个数值时，"值"也通常不译，如：

中文原文	啰唆或不当译法	简洁译法
pH 值	pH value	pH
平均值	mean value	mean
比值	ratio value	ratio
最大值/最小值	maximum value / minimum value	maximum / minimum 或 max / min
极限值	limit value	limit
峰值	peak value	peak
厚度值	thickness value	thickness

四、其他省译情况

从本质上讲，英文的省译，实际上就是省掉某些非关键成分后，不至于对原文意思造成误解，或者至少在特定语境下不造成误解。比如：

中文原文	不佳译法	简洁译法
棒材/线材/挤型材	bar materials / wire materials / extruding materials	bars / wires / extrusions
铸件/锻件	casting pieces / forging pieces	castings / forgings
压机/织机	pressing machine / knitting machine	press / knitter
点火器	flame igniter	igniter
漆包线绕线组	enamel winding wire set	enamel winding set

（续表）

中文原文	不佳译法	简洁译法
几十种微量元素和超微量元素	scores of kinds of microelements and ultramicro-elements	scores of microelements and ultramicro-elements
处于非标准状态下的气体	gas not under standard status	non-standard gas

以上这些例子，之所以能够省译，是简洁译法中的某个词已经包含了被省掉的词的意思。而"几十种微量元素和超微量元素"里的"种"，则属于英文里通常不译的量词。

五、使用后置或前置定语

例2.8：

原文：在英国，每六个成人里面就有一个已被诊断出患有常见心理疾病，如抑郁症或焦虑症。然而至少三分之二<u>被确诊者</u>未针对自己的问题接受治疗。

原译：One in six adults in England has been confirmed with a common mental disorder such as depression or anxiety. Yet at least two-thirds of <u>people that has been confirmed</u> receive no treatment for their condition.

改译：One in six adults in England has been confirmed with a common mental disorder such as depression or anxiety. Yet at least two-thirds of <u>people confirmed</u> receive no treatment for their condition.

分析：本例中的"被确诊者"原译采用定语从句，像这种简单的定语从句通常可以用过去分词当后置定语代替。

类似用法还有：

<u>所发现</u>问题	problems <u>found</u>
<u>所采取</u>方法	measures <u>taken</u>
<u>投保</u>财产	properties <u>insured</u>
<u>记录在册</u>产品	products <u>recorded</u>

但是，这种后置定语，很多都可以移到前头当前置定语，例如：confirmed people, recorded products。

例2.9：

原文：带法兰塑料弯头的两端焊接着法兰，可用于连接固定管道的塑料管件。

原译：The plastic bend with flanges has a flange at both ends, and can be used to connect plastic fittings of fixed pipes.

改译：The flanged plastic bend has a flange at both ends, and can be used to connect plastic fittings of fixed pipes.

分析：本句的"带法兰塑料弯头"译成plastic bend with flanges也没错，但译成flanged plastic bend显得更简洁明快，也更易理解。类似的表达还有valved port（带阀门接口）、reduced pipe（异径管）。

例2.10：

原文：钢材的待涂布表面应保持清洁干燥。

原译：The surface that will be coated of steels shall be kept clean and dry.

改译：The to-be-coated surface of steels shall be kept clean and dry.

分析：本例用组合词to-be-coated当前置定语，代替that will be coated，总体上显得更为简洁。其中to-be-coated surface也可表达成surface to-be-coated，将to-be-coated当作后置定语。

六、用单词替代从句或短语

例2.11：

原文：它是运用激光引擎技术所研发的第一代对人眼和身体无伤害的激光成像产品。

原译：It is the first generation of laser imaging product which does no harm to human eyes and bodies developed with laser engine technology.

改译：It is the first generation of laser imaging product harmless to human eyes and bodies developed with laser engine technology.

分析：本例将which does no harm to省译成harmless to，意思完全不变，但却更直接简短。这种将具有先行词的定语从句简化为无先行词且无助动词的定语形式，本质上意思不变，却更为简洁，是英语中经常采用的做法。

例2.12：

原文：过滤器主体与换网液压装置<u>以液压方式连接</u>。

原译：The filter body is <u>connected in hydraulic manner</u> to the screen changer hydraulic unit.

改译：The filter body is <u>hydraulically connected</u> to the screen changer hydraulic unit.

分析：本例用hydraulically替代in hydraulic manner，不仅更简洁，而且更符合英文行文习惯。

例2.13：

原文：将O形圈和垫圈放在无螺纹段。

原译：The O-ring and the washer are put on <u>a section without thread</u>.

改译：The O-ring and the washer are put on <u>a threadless section</u>.

分析：中译英时，碰到"无螺纹段"等"无***"的东西这样的名词时，可以使用***-less的定语，这样更简洁，读起来也更容易理解。

小结：在中译英时，对于表示动作方式的情况如果按字面意思翻译，会导致译文别扭费解。这时候，如果将原文换个说法，比如将"以液压方式连接"转换成英文副词+动词"液压地连接"（hydraulically connect），将"无螺纹段"的"无螺纹"理解成一个词而不是一个短语，再英译成threadless，就是符合英文表达的句式。其中，hydraulically是直接将形容词hydraulic变成副词。形容词转化成副词的方式，通常都是后面加-ly或-ally之类的，规定比较单一。而表示否定的英文前后缀则有很多，常见的表否定的前后缀如下表所示：

类别	词缀	举例
表否定的词缀	a-（不、无）	asymmetric, atypical
	anti-（反对、相反）	antibody, antiageing
	counter-（反对、相反）	counterclockwise, counteractive, countermeasure
	de-（非、相反、除去、向下）	deforest, deform, demerit
	dis-（不、无、相反）	disappear, disapprove, discolor, discontinue

（续表）

类别	词缀	举例
表否定的词缀	in- / im- / il- / ir-（不、无、非）	incorrect, inability, inaccurate impossible, impolite, imprudent illegal, illiterate, illogical irregular, irresistable, irresolvable
	mal-（不好、坏）	malfunction, malnutrition
	mis-（不、误）	misunderstand, misstatement, mislead, misfortune
	non-（不、无、非）	non-existence, non-productive, nonverbal
	un-（不、无、非）	unfair, unauthorized, unambiguous
	under-（不、欠、不充分）	underdeveloped, underground, undervoltage（欠压）
	-less	harmless, threadless, colorless, countless
	-free（无、不）	stress-free, cost-free
	-proof（防、抗）	waterproof, airproof, bombproof
表方式的副词	well-（良好的）	well-organized（组织良好的）
	ill-（不佳的）	ill-organized（组织不好的）

七、词性转换

在英文里，词的意思是很丰富的。有很多习惯上当名词用的词，其实也有动词的意思，而且如果把这些名词当动词用，会有意想不到的简洁生动效果。如：

例2.14：

原文：船上的家具通常用螺栓安装在甲板上。

原译：The furniture on a ship is often connected by bolts on the deck.

改译：The furniture on a ship is often bolted on the deck.

分析：本句原译没有任何错误，但是把be connected by bolts改译成be bolted后更简洁，更符合英文行文习惯。

同理，"液位计应以法兰连接"的译法Level gauges shall be connected with flanges也明显不如以下译法好：Level gauges shall be flanged。

例2.15：

原文：因为制动液会损坏油漆或树脂表面，所以小心不要把它溅到此类材料上。

原译：Be careful not to splash brake fluid on paint or resin surfaces as the brake fluid may cause damage to the surfaces.

改译：Be careful not to splash brake fluid on paint or resin surfaces as it may damage the surfaces.

分析：本例的"制动液会损坏油漆或树脂表面"，原译是the brake fluid may cause damage to the surfaces，改译后是it may damage the surfaces，两个译法意思都很清楚。但原译将damage当名词用，前面加一个使动词cause，而改译则将damage直接当动词用，改译后明显更干净利落。像这种能用一个动词表达的，一般就不要用使动词加名词性动词表达。

例2.16：

原文：一些发展迅速的高度工业化国家被称为新兴工业化国家。

原译：Some countries that are rapidly developing and highly industrialised are called Newly Industrialised Countries (NIC).

改译：Some rapidly developing and highly industrialised countries are termed Newly Industrialised Countries (NIC).

分析：本例将"被称为"的原译are called改成are termed，即把term当动词用，显得更简洁。term当动词用的情况并不多见，但在这里用得很恰当，让整个句子都显得更有活力。

例2.17：

原文：Cradles shall be situated within metallic bunks attached to the trailer and contoured to the loaded pipe diameter.

译文：托架应位于拖车的附带金属框架中，而且托架形状与接触处钢管的外径面一致。

分析：本句的contour平常都当名词用，这里却当作动词用，意思是"勾画……的轮廓"或"轮廓应成为什么形状"类似。

小结：英文与中文一样，都广泛存在一词多义的现象。中文有名词动用现象（如："国将不国""驴不胜怒，蹄之"），英文也有很多名词动用现象，但很多译员都未意识到名词动用的必要性，不了解名词动用的好处。下表列出了应用类文章里经常可以名词动用的一些词：

单词	正常意思	词性转换后意思
thread	螺纹	以螺纹连接
flange	法兰	以法兰连接
screw	螺丝	上螺丝
bolt	螺栓	栓接，以螺栓连接，上螺栓
bridge	桥	跨接，桥接
chamfer	倒角、削角	倒角、削角
case	箱、盒	装入箱（盒）内
bush	衬套	装入套（筒）内
lever	杠杆	撬开，起杠杆作用，使用杠杆
size	尺寸	使尺寸成为，确定尺寸，使尺寸足以……
term	术语	将……称为
factor	系数、因素	把……系数/因素考虑在内（在确定意思时，可译成"乘以系数"或"除以系数"）
power	功率、电源	供能、供电
label	标签	贴标签于
pair	（一）对、双、付	组对、使成双
man	人	操纵、给……配置人员，（manned：有人操纵的、驻人的）
round	圆的	使圆（round off：使圆满、弄圆、四舍五入）
target	目标	（target at）把……作为目标、瞄准
group	组、群	分组、分群
grade	等级、级别	分级、分等
top	顶、顶部	达到……的顶端，处于……的最前，超越
condition	条件	调节
back	背面、后面、背部	支持
shield	罩、盾	保护、罩住、屏蔽
guard	警卫、防备、保护	保护、防护
purpose	目的、用途、意志	决心、企图（purposed：打算的、计划的，purposedly：有目的地、蓄意地、故意地）

八、省略共同成分

例2.18：

原文：工艺中采用的方法不一致，或特定的程序被旁路，都可能影响系统完整性和可靠性。

原译：Inconsistent methods may be applied or specific procedures <u>may be</u> by-passed in the process, which may compromise system integrity and reliability.

改译：Inconsistent methods may be applied or specific procedures by-passed in the process, which may compromise system integrity and reliability.

分析：本句改译后，specific procedures后面省略了may be，这是一个典型的省略共同成分的例子。

例2.19：

原文：其区别几乎可以一眼看出，因为蜘蛛总是有八条腿，而昆虫的腿不会超过六条。

原译：One can tell the difference almost at a glance, for a spider always has eight legs and an insect <u>has never</u> more than six.

改译：One can tell the difference almost at a glance, for a spider always has eight legs and an insect <u>never</u> more than six.

分析：本句改译后省略了前一句出现过的has。另外，还需要说明一个相关知识点，即蜘蛛不是昆虫。（昆虫的定义是：身体分为头、胸、腹三段，一生形态多变化，成虫通常有2对翅膀6只足；头上有1对触角，骨骼包在体外部。）

九、说法变换或句式转换

由于中文与英文的显著差异性，中文的很多词语或者句式都不能字面化翻译，尤其是不能逐字翻译，而是要经过一定的变换或转换。请看以下例子：

例2.20：

原文：配备了螺旋管道加热器，可定时自动对出料管道内的药剂加热，从而提高送料精度并<u>防止药剂受潮</u>。

原译：A spiral tube heater is added to automatically heat reagent in the discharge pipe in a timely manner, thereby improving the feeding accuracy and preventing the reagent from being affected with damp.

改译：A spiral tube heater is added to automatically heat reagent in the discharge pipe in specified time, thereby improving the feeding accuracy and preventing wetting of the reagent.

分析：对于"防止药剂受潮"这样的译法，多数译员都倾向于用被动译法。但如果换个说法，采用主动式，会让句式简洁很多，而且更符合英文行文习惯。

例2.21：

原文：当气缸中充入压缩空气时，径向柱塞所产生的推力，通过横梁及滚轮与内曲线导轨相互作用，产生扭矩输出，经行星减速机构减速产生强大扭矩，驱动闸阀的阀杆螺母旋转，使阀杆作升降直线运动，开启或关闭闸阀。

原译：When the cylinder is filled with compressed air, the radial plunger produces a thrust. Due to interaction of the beam, roller and inner-curve guide rail, the torque output is produced. After the speed is reduced by the planetary reducer, large torque is produced to drive the stem nut of the gate valve to rotate, so that the stem linearly moves up and down to open or close the gate valve.

改译：When the cylinder is filled with compressed air, the radial plunger produces a thrust, outputs a torque due to interaction amongst the beam, roller and inner-curve guide rail, generating a strong torque after the reduction of the planetary reducer, which further rotates the stem nuts of the gate valve, moves the stem linearly up and down to open or close the gate valve.

分析：本例中"驱动闸阀的阀杆螺母旋转"里的"驱动……旋转"，实际上就是"使旋转"的意思，可以直接译成rotate，而不是译成drive ... to rotate。

例2.22：

原文：另外，由于本系统是基于Salve-Master的物理架构来进行任务分配及调度的，因此与目前市面上其他同类相比，本系统还满足了多用户同时使用的需求。

原译：In addition, the system carries out task assignment and deployment on the basis of the physical structure of Salve-Master, therefore, compared with the similar systems available in the market, the system also can meet the requirements that allow multiple users to use it simultaneously.

改译：Moreover, the system assign and deploy tasks based on the Salve-Master physical structure, therefore, compared with similar systems in current market, the system also allows multiple users to use it simultaneously.

分析：本句中的"基于Salve-Master的物理架构来进行任务分配及调度的"，其意思就是"基于Salve-Master的物理架构分配及调度任务"，"来进行……的"是中文里非常常见的一种半口语式表达。这种半口语式表达的动作，一般不宜译成carry out task assignment and deployment这样的carry out + 名词化动词的形式，而是要直接把assign和deploy当作句子的谓语动词。

另外，本句最后的"本系统还满足了多用户同时使用的需求"，原译顺着原文的句式走，译成the system also can meet the requirements that allow multiple users to use it simultaneously，这样译基本没什么问题。但如果把本句译成the system also allows multiple users to use it simultaneously，会更简洁明了。

十、不啰唆、不绕弯翻译

在做翻译时，若感觉原文句子不够简单明了，有点绕，甚至不得不用解释性翻译时，通常就是碰到盲点了，这时候，要多查词典、网络，如果还觉得有问题，可以问一下有经验的翻译人员。请见以下例子：

例2.23：

原文：The sound pressure level and sound power level values for your machine are provided in the tables shown in 13.4.

原译：您机器的声压级值与声能级值均可参见第13.4中所示表格。

改译：机器的声压级与声能级如第13.4中表格所示。

分析：应用类文件里的人名或人称代词，只要不是很必要的，就可以省略掉。另外，are provided in the tables shown in 13.4译成"均可参见第13.4中所示表格"不如译成"如第13.4中表格所示"简洁。也就是说，原文啰唆时，译文不要跟着啰唆，而且在确保表达出原文意思的情况下，可以变换说法。

例2.24：

原文：测量由测量班对路面中心线及边线的位置和高程进行复测。

原译：The measurement is measured by the measurement class at the location and elevation of the center line and the side line of the road.

改译：It is measured by the measurement group at the location and elevation of the center line and the side line of the road.

分析：本句原文里出现了三个"测"字，感觉上有点啰唆。所以，译文不宜把三个"测"都直译出来，而是根据语境进行翻译。

例2.25：

原文：换热器的表面必须经过去除毛刺表面处理。换热效率确保达到80%。

原译：The surface of the heat exchanger must be subject to deburring surface treatment. The heat exchange efficiency shall reach 80%.

改译：The surface of the heat exchanger must be deburred. The heat exchange efficiency shall reach 80%.

分析：本句的"经过去除毛刺表面处理"其实就是"去毛刺"，这是典型的中文特色表达，译成be subject to deburring surface treatment后，看似很准确，但实际上却会让英语母语者觉得很绕，莫名其妙。

以下是另一个类似例子：

例2.26：

原文：所有使用到的非不锈钢材质的型材表面都需进行热镀锌防腐处理，采用热镀锌。

原译：The surfaces of all stainless steel sections employed shall be subject to hot dip galvanized anti-corrosion treatment, and hot dip galvanization shall be adopted.

改译：The surfaces of all stainless steel sections employed shall be hot dip galvanized for anti-corrosion.

分析：本句的"进行热镀锌防腐处理，采用热镀锌"的意思，其实就是"为防腐而热镀锌"，其中"进行"是常见的中文辅助动词，无实质意义。后面的"采用热镀锌"则是重复的表达，也不必译出。

十一、适当多用短句，少用长句和复杂句

例2.27：

原文：Temporary rates paid to an employee are monitored for possible abuse or the need for a permanent job reclassification.

The payroll department is promptly and formally notified when employees are terminated or transferred and when other payroll changes are made.

The pay histories of transferred employees are updated so that year-to-date figures reflected under the old and new work location codes (e.g., CISCO) are correct.

译文：对支付给员工的临时款项进行监测防止违规行为并及时发现需要进行永久性工作重新分类的情况。

如果有员工终止雇员关系或更换职位或有其他工资变动时需及时向工资管理层进行正式通报。

对转岗员工的工资记录进行更新以确保新旧工作地代码（如CISCO）下反映的年初至今数字均准确无误。

改译：对支付给员工的临时款项进行监测，防止违规行为，并及时发现需要进行永久性工作重新分类的情况。

如果有员工终止雇员关系或更换职位，或有其他工资变动时，需及时向工资管理层进行正式通报。

对转岗员工的工资记录进行更新，以确保新旧工作地代码（如CISCO）下反映的年初至今数字均准确无误。

分析：本例的译文总体上没有什么错误，唯一的问题是句子太长。在实用翻译中，不仅要求对原文的理解和表达正确，而且要求翻译后的句子容易理解，其中使用短句，少用长句和复杂句就是最重要的一个要求。这一要求，也是纯文档编写（非翻译）的一个重点要求。

十二、图片、表格内容的简化译法

图片和表格的内容,不论是标题、表头还是正文,一般都要求译得很简洁,有点类似于标题的翻译。而且与一般标题不同的是,有些图片和表格空间较小,但要求表达的信息量却相对较大,因此通常译得比普通标题更加简洁,普通标题通常只使用很常见的缩写词,但图片和表格却会使用业内常见,但非业内人士看不懂的缩写。这里暂不涉及图片和表格的英译中,只介绍中译英可使用的一些简化译法。

例2.28:

原文:

汽轮机VWO工况下各级抽汽参数表

抽汽级数	流量 kg/h	压力 MPa(a)	温度 ℃	允许的最大抽汽量 kg/h
第一级(至1号高加)	140173	7.632	385.9	
第二级(至2号高加)	167832	5.021	328.5	
第三级(至3号高加)	80048	2.411	468	
第四级(至除氧器)	101636	1.207	366.4	
第四级(至给水泵汽轮机)	102452	1.207	366.4	
第四级(至厂用汽)				70000
第五级(至5号低加)	53057	0.409	232.2	
第六级(至6号低加)	51237	0.222	166.5	
第七级(至7号低加)	51102	0.112	102.8	
第八级(至8号低加)	87851	0.05	81.5	

汽轮机发电机组临界转速表(按轴系、轴段分别填写)

轴段名称	一阶临界转速 r/min		二阶临界转速 r/min	
	轴系	轴段	轴系	轴段
高中压转子	1692	1650	>4000	>4000
中压转子	/	/	/	/
低压转子 I	1724	1670	>4000	>4000
低压转子 II	1743	1697	>4000	>4000
发电机转子	984	933	2676	2695

译文：

Extraction parameters at all levels under VWO condition of turbine

Extraction stage	Flow kg/h	Pressure MPa (a)	Temp ℃	Max allowable extraction kg/h
Stage I (to No. 1 HP heater)	140173	7.632	385.9	
Stage II (to No. 2 HP heater)	167832	5.021	328.5	
Stage III (to No. 3 HP heater)	80048	2.411	468	
Stage IV (to deaerator)	101636	1.207	366.4	
Stage IV (to feed pump turbine)	102452	1.207	366.4	
Stage IV (to auxiliary steam)				70000
Stage V (to No. 5 LP heater)	53057	0.409	232.2	
Stage VI (to No. 6 LP heater)	51237	0.222	166.5	
Stage VII (to No. 7 LP heater)	51102	0.112	102.8	
Stage VIII (to No. 8 LP heater)	87851	0.05	81.5	

Critical speed of turbine generator set (fill in by shaft system and shaft segment respectively)

Name of shaft segment	1st critical speed r/min		2nd critical speed r/min	
	Shaft system	Shaft segment	Shaft system	Shaft segment
HP and MP rotors	1692	1650	>4000	>4000
MP rotor	/	/	/	/
LP rotor I	1724	1670	>4000	>4000
LP rotor II	1743	1697	>4000	>4000
Generator rotor	984	933	2676	2695

分析：以上译文中加下画线的都是缩写。表格中译英的注意点：文章中使用表格的目的，通常是把比较复杂的事情直观显示出来。表格里的信息量通常比较大，而表格的分列式表达法，又让格子的横向空间显得越发有限。再加上英文通常比中文更长，使格子空间显得更加拥挤。所以在表格中译英时，需要尽可能地简化译文。

简化译文的方法，除了前文提到的方法之外，还包括将标题的"表"或"图"的省略不译，相应的，在英译中时，则通常可以用增译法，把不含有figure、table的图表标题译成"……图""……表"。

机械工程类文件所含表格里，有些词通常可以使用英文缩写，下表所列只是一小部分例子：

中文	English	中文	English
高温 / 低温 / 中温	HT / LT / MT	高压 / 低压 / 中压	HP / LP / MP
高温高压	HTHP	温度	Temp / T
直流 / 交流	DC / AC	数量 / 质量	Qty / Qlty
制造商	MFR	软钢 / 碳钢 / 不锈钢	MS / CS / SS
最大（值）/ 最小（值）	Max / Min	紫外线 / 红外线	UV / IR

本节课后练习

翻译下列句子：

1. 新研究预计海平面下世纪将大幅升高
2. 三位美国经济学家荣获2013年诺贝尔经济学奖
3. 品质保证部
4. 中国工商银行财务报表
5. 增长型行业
6. 敏感性皮肤
7. 催化裂化法制丙烯
8. 特种钢
9. 隐性基因
10. 非典型性肺炎
11. 电压输入型电流钳
12. 电冰箱
13. 雷诺准数值
14. 研磨机
15. 被转让资产

16. 无应力配管
17. 在生物学上，狮子和老虎同属于猫科动物。
18. 在原始文档级别格式中，每个开始标记与一个结束标记配对。
19. 这张椅子的腿是用螺丝固定在地板上的。
20. 这台机器噪声很大，该上润滑油了。
21. 本文着重描述了这些方法的研究现状及其特点，并对边坡稳定性分析方法的发展做了展望。
22. Level gauges shall have flanged connections with flange facing and rating in accordance with the appropriate Process Piping Specification.
23. Accounts receivable records are promptly updated to reflect payments received.

第二节 时态、语态和句式

在做中英翻译时，时态、语态及句式的选择是很重要的，而且，往往确定了时态和语态，就基本确定了句式。如果时态语态或句式选择不当，轻则让人感觉句子不顺畅，重则无法理解。而在做英汉互译时，由于中文不强调时态和语态，因此重点介绍中文译文的句式。

一、中译英

在做中译英时，通常而言，语态的选择根据动作的性质确定，时态的选择根据动作发生的时间和语境确定。主动发生的事情通常用主动语态，被动发生的事情通常用被动语态。但是，对于不强调或不必说明施动者的句子，通常会把受动者当作主语，译成被动。而时态则一般用现在时或过去时，只有在强调动作已经完成的情况下才用完成时。

例2.29：
原文：

故障	原因	对策
调直筒振动	调直轮支架调整偏重	把5套调直轮全部松到底重新受力调整

原译：

Trouble	Cause	Countermeasures
Straightening tube vibrates	The straightening wheel bracket is too heavy.	All five sets of straightening wheels are loosened completely and go through stress adjustment.

改译：

Trouble	Cause	Countermeasures
Straightening drum vibrates	The straightening wheel bracket is over regulated.	Loosen all five straightening wheels completely and re-regulate the force applied.

分析：本例是典型的机械类说明书的一部分。对于这类句子，通常都可以按照一定的方法选择语态、时态和句式。

具体到本例中，其中故障说明部分，因为"调直筒振动"是已经出现的问题点，而且是调直筒本身自动振动，所以用主动语态现在时。原因部分，"调直轮支架调整偏重"是指支架目前被调得过重，因此用被动语态现在时。而对策部分，是要求对策人员做出的、尚未发生的动作，是命令、要求式中文祈使句，因此要用祈使句现在时。通常而言，对策以及操作指令等，由于动作尚未发生，是要求操作人员完成的，一般用祈使句。

但要注意，语态和时态并不是固定的，具体要根据上下文确定，有时主动或被动语态都可行，或者可采用多种句式。比如，在用英文祈使句翻译对策部分时，既可以用无主语式祈使句，又可以用 should、shall、must 引导的祈使句。

例2.30：

原文：如测试探针脏或位置发生偏移，导致接触不良，则使用带酒精的棉棒进行清洁，调整探针的位置，确保探针接触良好。

译文：If the testing probe is polluted or shifted, and results in poor contact, clean it with an alcohol swab, and regulate the position of the probe, to ensure a good contact.

分析：本句是比较典型的能直接顺着原文句式翻译的句子，其中后几句都用祈使句。

例2.31：

原文：<u>左击</u>变量文本，打开的弹出菜单<u>显示有</u>变量文本的对应类型选项。

原译：<u>Left-click</u> the variable text, the pop-up menu opened up <u>will display</u> has a corresponding type option to the variable text.

改译：<u>Left-click</u> the variable text, and the pop-up menu opened <u>will show</u> the type option of the variable text.

分析：本句是一个承接复合句，前一分句是一个中文祈使句，这里直接译成英文祈使句。后一分句承接前一分句，说明前一分句的结果，因此用and表示承接关系。其中，原文的"显示有"就是"显示"，"有"不必翻译。

例2.32：

原文：<u>出现异常情况</u>，如设备中有异物，设备异响，局部变形，<u>安全装置破损</u>等，<u>应立</u>即停止运转，相关人员撤离操作场所及周围环境；报请现场主管人员，请专业人员查找原因进行修理。

原译：If <u>there were</u> some exceptions, for example, foreign matters in the equipment, abnormal noise of the equipment, local deformation, <u>safety devices were damaged</u>, <u>we should</u> stop the operation immediately, the workers evacuate from the operation place and the surroundings; report to the supervisor on the spot, ask the professionals to find the reasons and repair it.

改译：<u>In case of</u> any abnormity, for example, foreign matters in, abnormal noise, local deformation of the equipment, <u>damages to safety devices</u>, the operation <u>should be</u> stopped, workers <u>should be</u> evacuated from the site and the surroundings immediately; the conditions should be reported to onsite supervisor, and professionals should be asked to find the causes and repair it.

分析：本例除了数处用词不当，即"异常"—exceptions应改成abnormity，故障"原因"—reason应改成cause外，主要的问题是语态和句式选择不当。具体如下：

（1）there be句式：原译中所用的这种句式，原则上并无错误，但是

there be句式在英文里用得并不多，在应用类翻译中也用得较少。当表示"出现"某些意外情况时，通常可用in case of + n.结构。

（2）在本句中，将"出现异常情况"译成in case of + n.结构后，后面的举例表达成名词结构应该比句子结构更合理。原译使用了三个名词结构，但第四个safety devices were damaged不是名词结构，与前面几个名词式表达不协调，宜改译成damages to safety devices。

（3）we should句式：在应用类文件中，尤其是说明类文件中，如果原文未出现人，通常不使用人（名词或代词）做主语表示施动者，而是把受动者或其他相关词当作主语，用被动语态表达。

（4）原文"应立即停止运转，相关人员撤离操作场所及周围环境"里的两个小分句，中文句式看起来不一致，但实际意思都是提出要求，因此，在英译时，采用相同句式是一种较好的选择。

（5）原文最后部分"报请现场主管人员，请专业人员查找原因进行修理"，原译使用了英文祈使句。但在改译语境下，译成被动句会显得与前面的译文更协调。

例2.33：

原文：作业时必须按照安全工作制度要求落实安全防护措施，做好安全防护工作，严禁作业前喝酒或带病工作。

原译：During the operation, make sure the implementation of safety protection measures in line with the safe work system and properly do the work of safety protection. Operation after drinking or while sick is strictly prohibited.

改译：During the operation, safety measures must be implemented in line with the safety work system. Operation after drinking or while sick is strictly prohibited.

分析：本段的原文，都是发出指令的祈使句。原译将第一句译成祈使句，第二句用be prohibited的句式，但第一句略显啰唆冗长。其中，

（1）make sure the implementation of safety protection measures ... 这样

的译法，不如safety measures must be implemented ... 这样的表达简洁干脆，make sure的语气也没must强。

（2）原文"落实安全防护措施，做好安全防护工作"意思重复，这是中文很常见的重复表达，在译文中没必要全部译出，可以省译后面一部分。

这种根据上下文选择句式的技巧以及省译的方法，都是译员必须掌握的。

例2.34：

原文：整套装置内设有流量计、液位和料位传感器，在液位和料位低时输出报警信号，防止设备和使用现场出现不必要的损坏和损失。

译文：The device is provided with a flowmeter, a liquid level sensor and a material level sensor which give alarm signals in low liquid or material level, to prevent unexpected damage to and loss of the equipment and the operating site.

分析：本句中需要说明的注意点有：

（1）"整套装置内设有流量计、液位和料位传感器"是一个中文主谓句，指的是这个装置里设有流量计、液位和料位传感器，由于施动者不明确或无需说明，可将受动对象当作主语，译成被动式the device is provided with ... 其中，"整套装置"可以只译成the device，不必译出"整套"。

（2）"在液位和料位低时输出报警信号"的主语是"液位和料位传感器"，因此，整个句子相当于"液位和料位传感器在液位和料位低时输出报警信号"，在中文语法中，属于主动式主谓句。这种中文主动式主谓句，由于施动者很明确，通常译成英文的主动句，但为避免重复，使用which当句子的主语，引导定语从句。其中，"液位和料位传感器"是指液位传感器和料位传感器，是复数的，因此give用原形。

例2.35：

原文：焊接曾储存易燃、易爆物品的容器时，焊前要用蒸气或碱水冲洗干净，并打开容器的所有孔口，经检测确认安全后，方可实施焊接。

译文：The containers which once stored the inflammables and explosives should be welded only after being cleaned by steam or soda water, with all container orifices opened and the safety ensured by detection.

分析：本句需要说明的注意点有：

(1) 本句的原文，实际上相当于以下句子：

在焊接……容器时，（焊工在）焊前要用蒸气或碱水（将容器）冲洗干净，并打开容器的所有孔口，经检测确认安全后，（焊工）方可实施焊接。

因此，本句原文其实是省略主语的数个主谓句。

这种省略主语的中文主谓句，与前例一样，都是施动者不重要，没必要说明的情况。因此，也要把受动对象（容器）当作主语，采用被动句。如果硬找出一个主语译成英文，译文会显得很生硬。

(2) "曾储存……"译成从句which once stored...是常见的译法。"焊前要用蒸气或碱水冲洗干净"译成should be welded only after being cleaned by steam or soda water的句式，则是转换说法，是译员需掌握的技巧。

(3) 原文第一小句是时间状语，其受动对象是"容器"，与第二小句的受动对象相同，在本句的语境下，可以把时间状语转化成非时间状语，变成定语从句限定"容器"，并把"容器"作为前两个小句的主语。具体看参考译文。

当然，如果把第一小句译成时间状语，也是可以的：

The containers should be cleaned by steam or soda water before being welded if the inflammables and explosives have been stored in them before, and all container orifices should be opened and the safety should be ensured by detection.

但是，如果采用以上译法，感觉上会觉得没有参考译文那么紧凑。

例2.36：

原文：（管道中）配备了螺旋管道加热器，可定时自动对出料管道内的药剂加热，从而提高送料精度并防止药剂受潮。

原译：A spiral tube heater is added to automatically heat the reagent in the discharge pipe in time, improving the feeding accuracy and preventing the reagent from being affected with damp.

改译：A spiral tube heater is added to automatically heat the reagent in the discharge pipe in time, improving the feeding accuracy and preventing wetting of the reagent.

分析：原译将"防止药剂受潮"译成被动式preventing the reagent from being affected with damp，是很中式英语的表达，应译成主动式preventing wetting of the reagent。

例2.37：

原文：广义的跨境电子商务是指分属不同关境的交易主体通过电子商务手段达成交易的跨境进出口贸易活动。

原译：Broadly, the cross-border e-commerce refers to the cross-border import and export trade activities in which the transactions are concluded by the transaction participants who belong to different custom territories respectively by means of e-commerce.

改译一：Broadly, the cross-border e-commerce refers to the cross-border import and export trade activities in which the transactions are concluded by the transaction participants belonging to different custom territories respectively by means of e-commerce.

改译二：Broadly, the cross-border e-commerce refers to the cross-border import and export trade activities in which the transactions are concluded by the transaction participants in different custom territories respectively by means of e-commerce.

分析：本句原译文并没有明显问题。这里要说明的是定语（从句）的译法。这里的"分属不同关境的"译成 who belong to different custom territories 不如译成 belonging to different customs territories 简洁，更不如 in different custom territories 直接。

以后中译英时，在合适的地方，可以适当少用定语从句，以简化句式，比如以上改译一的译法。有时候还可以在不改变原文意思的情况下，用更简洁的说法表达，如在本例中，将"分属"理解成"位于"，译成 in，就更简单了。

实际上，大家只要多加思考，就会发现，在英文里，尤其是应用类英文里，使用带先行词的定语从句的比例并不是很高，尤其是在直接用简单定语就能表达清楚的情况下，一般可以不用更麻烦的定语从句。推而广之，能简单表达的，一般不要用麻烦的表达法。毕竟，应用类文件的目的只是为了传递信息，在不损失所传递的信息的情况下，能更简洁明了地表达原文意思，何乐而不为呢？

例2.38：

原文：为配合社保管理规定，员工入职或离职在当月10日之前的，当月办理缴纳或停缴社保及公积金；在此日期之后入、离职人员，均在下月给予办理社保公积金事宜。

原译：To comply with the social security management regulations, the social security and reserved funds of the staff who induct into or leave the office before 10th day of the month shall be paid or stopped being paid on the month; and who induct into or leave the office after 10th day of the month shall be paid or stopped being paid on the next month.

改译：In accordance with the social security management regulations, the payment of social security and reserve funds shall be begun or suspended that month for the employees who join or leave the office before the 10th day of the month, or begun or suspended the next month for the employees who join or leave the office after that day.

分析：译好本句的关键，是要把句子的公共成分（"社保公积金"的缴纳或停缴）当作主语，把句子的类似但不同的成分（"员工入职或离职在当月10日之前"和"在此日期之后入、离职人员"）当作目的状语。

二、英译中

对于中文基础不太差的译员，一般都会觉得英译中比较容易切入。但实际上，英译中文件，不仅会碰到难以理解的原文导致无法准确翻译，而且会碰到理解原文但无法准确或顺畅地表达成中文的情况。

另外，如前所述，由于中文不强调时态和语态，因此本部分主要介绍英译中时译文的句式。请看以下例子。

例2.39：

原文：The use of water at a temperature not exceeding 30℃ <u>is designed to</u> limit the temperature rise in the grout during the grouting operation.

原译：水温设为30℃以下<u>是为了</u>限制灌浆过程中浆液温度升高。

改译：使用30℃以下的水的目的是为了限制灌浆过程中浆液温度升高。

分析：原译把"水温设为30℃以下"作为主语，意思表达不够明了。将主语改成"使用30℃以下的水的目的"之后，整个句子结构更加清晰。

例2.40：

原文：All such items as pipes, fittings and valves, <u>whether or not specifically mentioned or indicated there</u>, shall be furnished and installed <u>if necessary to</u> complete the system in accordance with the best practices of the plumbing trade and to the satisfaction of the Employer's Representative.

原译：<u>如果必要</u>，管道、设备和阀门等所有项目，<u>无论这里是否被提及或提示</u>，都应被提供和安装，以完成符合水暖工作的最佳实践和雇主方代表的满意度的系统。

改译：<u>为按照水暖行业的最佳做法，以业主代表满意的方式做好整个系统，如有必要，无论是否明确提及或指示，均应提供和安装管道、配件和阀门等。</u>

分析：本句原文后半部分有个if引导的条件状语和to引导的目的状语。在英文里，假设状语（或从句）和目的状语（或从句）经常放在句子后头，但在中文里，假设状语和目的状语更多的是放在前头。因此，本句改译后，把两个状语从句都提前，并对句子做相应调整。

例2.41：

原文：Each load of material shall be placed in the embankment to produce the best practicable distribution of the material. The materials shall be placed <u>so that</u> the permeability will gradually increase towards the downstream boundary of the zone.

Fill adjacent to a concrete structure shall be placed and compacted with such equipment and in such manner that no damage is caused to the concrete structure. The placement of fill material shall be deferred until the concrete structures have attained an age of 28 days as directed by the Engineer.

译文：各批物料在路堤中铺设的位置应可以使物料达到尽可能合理的分布。物料的铺设方式应使得其渗透性朝着该区域的下游边界逐渐增加。铺设和压实与混凝土结构相邻的填充料时所使用的设备以及设备的使用方式，不得对混凝土结构造成损害。填充料的铺设时间应往后推迟至混凝土结构达到工程师指示的28天日龄时。

分析：本例第一段第二句的so that，表示的是place的方式，因此，如果把这一句译成"应铺设材料，以使得……"就完全弄错意思了。第二段的such和such manner that也是类似的用法。因此，参考译文将这两处译成"……方式应使得"和"……方式不得"。

本例的本质，还是英文与中文表达方式的不同。英文so that / such that句式的that后面跟的是目的状语从句。其中，The materials shall be placed so that …一句中，主语是the materials（物料），动词是be placed（铺设），而中文则译成"物料的铺设方式应使得……"，即把"（铺设）方式"当主语，"使得"当谓语，"其"当宾语，同时"其"还是后面的主语（即使用兼语句形式）。

例2.42：

原文：Either party may terminate the Agreement at any time by notice in writing to the other:

(i) if the other party is in breach of the Agreement, or the breach is not remedied within 30 days after it is required to be remedied; or

(ii) if the other party is in persistent breach of the Agreement, whether such breaches are materials or not or remedied or not.

原译：本协议任何一方均可在书面通知另一方后，于任何时候终止本协议：

（i）若另一方违反本协议规定，或在被要求采取补救措施的通知后，并未在30天内补救；或者

（ii）若另一方坚持违约，无论该等违约行为是否严重或是否可补救。

改译：在以下情况下，本协议任何一方均可在书面通知另一方后，于任何时候终止本协议：
(ⅰ) 另一方违反本协议规定，或在被要求采取补救措施后，未能在30天内补救；或
(ⅱ) 另一方持续违约，无论该等违约行为是否严重或是否可补救。

分析：本例是法律英语里典型的分点式条件句。对于这种句子，原译跟着原文句式走，虽然也表达出了原文的意思，但这样的句式不符合中文的行文习惯。中文的条件句经常提前，如果不提前，也要有个相应的说明，如改译后的说明"在以下情况下"，这样，就相当于把条件句提前了，这里姑且称之为"条件句假提前"。在英文中有多个选择性条件句并列时，经常可以使用"条件句假提前"的翻译方法。

但是，对于英文只有一个条件句的情况，如果也用中文的"条件句假提前"做法，则有点不妥，宜把条件句实际提前。如：

原文：Either party may terminate the Agreement at any time by notice in writing to the other if the other party is in breach of the Agreement, or the breach is not remedied within 30 days after it is required to be remedied.

原译：在以下情况下，本协议任何一方均可在书面通知另一方后，于任何时候终止本协议：另一方违反本协议规定，或在被要求采取补救措施后，未能在30天内补救。

改译：如另一方违反本协议规定，或在被要求采取补救措施后，未能在30天内补救，则本协议任何一方均可在书面通知另一方后，于任何时候终止本协议。

例2.43：

原文：In order to make the patterns comparable across countries, nominal sectoral shares (nominal value added of a sector in proportion to total nominal value added) are plotted against the level of economic development measured as the logarithm of per capita GDP in 1990 dollars.

原译：为了使模式在各国间具有可比性，可以将名义产业份额（产业名义附加值与名义附加值总额之间的比例）对经济发展水平作图，在图中表现为以1990年美元为基准的人均GDP的对数。

改译：为了使各国间具有模式的可比性，可以将名义产业份额（产业名义附加值与名义附加值总额之间的比例）对经济发展水平作图，其中经济发展水平折算成以1990年美元为单位的人均GDP的对数。

分析：本例原译已经表达出原文的意思，但measured as the logarithm of per capita GDP in 1990 dollars的译文没表达好。这里可以采用补充说明的形式译出。

例2.44：

原文：The prospect of life with little or no cash, at least for a few days, cheered those who think Indians should be switching to smartphone apps and card-based payments, which are easier for the authorities to track and tax.

原译：只有少量现金或者没有现金的生活景象（至少那几天是）鼓舞了那些认为印度应该使用手机应用程序和银行卡付款的人，他们觉得手机应用程序和银行卡付款更方便于当局追踪和收税。

改译：至少在一段日子内只有少量现金或者没有现金的生活景象，让那些认为印度会转而使用手机应用程序和银行卡付款，从而更有利于当局的追踪和税收的人们欢欣鼓舞。

分析：本句的背景是2016年年末印度总理宣布废除大额纸币。原译的句子顺序不大合理，整体表达不佳，不符合中文表达方式。修改后的句子虽然仍然是复杂句，但总体上显得更有条理，更符合中文行文习惯。

例2.45：

原文：Distribution transformers are a fundamental component of the electricity distribution network powering our national economies. By changing voltages, transformers lower voltages to levels that are needed by end users, they comprise part of the voltage transformation system enabling high-voltage power transmission and distribution (T&D) necessary to reduce overall network energy losses. Compared with other electrical equipment, transformers are very efficient, typically incurring losses of just 1%–3% when changing from one voltage level to another. However, the fact that nearly all electricity passes through transformers prior to final use means opportunities to reduce losses in distribution transformers are highly significant for improving the electricity network efficiency overall.

A network is predominantly comprised of the power lines that connect the power plant with the final point of demand; <u>however, in order to minimise losses and ensure safe usage, transformers are used to reduce the load current in the transmission line between the generating plant and the distribution network.</u>

译文：配电变压器是为国家经济发展提供电力的配电网络的基本组成部分。<u>变压器可变换电压，将电压降至终端用户需要的水平。变压器是电压变换系统的组成部分，可实现高压输配电，从而降低整体电网能耗。</u>与其他电气设备相比，变压器的能效较高，一般而言，在其电压等级发生变化时，能耗仅为1%—3%。然而，在最终用电前，几乎所有电力均需通过变压器，这意味着我们有机会降低配电变压器的能耗，这对提高总体电网效率具有十分重要的意义。

电网主要由输电线组成，可将电厂与最终电力需求端相连接；然而，为了尽量降低能耗并确保用电安全，<u>应使用变压器降低发电厂和配电网之间的输电线的负载电流。</u>

分析：本例加下画线部分如果顺着原文的句式译，就会显得很生硬。译文根据上下文背景，做了一定的句式调整。具体请看参考译文。

本节课后练习

翻译下列句子：

1. 当设备剧烈震动时，按下紧急开关按钮，以断开电源。
2. 当切割圆盘电机不能运转时，可能是由于气压不足，可调高气压。
3. 材料库房里严禁吸烟，库内电器设备必须完善，不得产生火花，要经常检查库房消防设施的完好情况，并熟悉使用消防设备，做好防火工作。
4. 若因使用本公司生产的医疗器械产品导致人员伤亡、财产或环境伤害等质量事故时，还应按国家、地方有关质量事故报告制度的要求，及时向有关政府管理部门进行事故报告。
5. 如果是采用试验车对试验路面洒水，则将贮水罐安装在牵引车上。洒水喷嘴应将水均匀地喷洒在试验轮胎将要通过的区域，且尽量少溅出水花。

6. The hydraulic stacker must be repaired when there is any:
   - Dent or damage;
   - Corrosion or oxidation;
   - Cracks in welds or structural components.
7. Pipes shall be stored under a shade, in such a manner as to allow air to circulate freely among the pipes and thereby avoid excessive build up of heat in the pipes.

第三节　行文表达

一、正式用语（避免口语化）

英文应用文翻译成中文时，倾向于用正式语体，但也有用非正式语体的。

序号	正式程度低	正式程度高	举例
1.	的	之（或省略）	正确的对策→正确对策
2.	（如果）……的话	如/若……	（如果）有必要的话，联系相关厂家→如/若有必要，联系相关厂家
3.	那么/就	则	如果证实这些因素中存在的实质性差异，那么该分析的可靠性也会受到影响。→如证实这些因素中存在的实质性差异，则该分析的可靠性也会受影响。
4.	用	以/凭	用这种工具操作→以该工具操作
5.	来（表目的时）	以（可视情况省略）	确定最佳方法来修复该问题→确定最佳方法（以）修复该问题
6.	了（表示完成）	（多数可省略）	该医疗器械的使用，有可能引起了健康危害，或者已经引起了健康危害。→该医疗器械的使用，可能引起健康危害，或者已经引起健康危害。
7.	这里	此处	安装方法基本相同，这里不再重复说明→安装方法基本相同，此处不再重复说明
8.	那里	处/之处/该处	从制造商那里得到帮助→从制造商处得到帮助 那里的产品→该处产品

（续表）

序号	正式程度低	正式程度高	举例
9.	我	本人	我在此再次证明以下所盖印章为公证员公章→ 本人在此再次证明以下所盖印章为公证员公章
10.	不能	不得/无法	焊缝不能有一丝缺陷→焊缝不得有任何缺陷 该机器韧性差，不能承受较大的冲击力→ 该机器韧性差，无法承受较大冲击力
11.	没有/没	无/未/未有	没达到图纸要求的要继续进行压制调整，直至达到图纸要求为止→ 未达到图纸要求的要继续进行压制调整，直至达到图纸要求为止 确认零部件没有缺失→确认零部件未有缺失
12.	还	尚	还需要进行润滑→尚需进行润滑 还没有安装完毕→尚未安装完毕
13.	是不是	是否	检查连接是不是牢固→检查连接是否牢固
14.	可以	可行/允许	如果可以的话，关掉开关→如果允许，关掉开关
15.	只有	仅/仅有	只有在机器停下后才可拆卸→仅在机器停下后方可拆卸
16.	才/才可	方/方可	
17.	和	与/及/以及	运行和维护→运行与维护 总经理、董事、监事、秘书和财务主管和其他高管→总经理、董事、监事、秘书和财务主管及其他高管
18.	人称代词	适当时省略	如果你有灭火技能，尝试灭火→如有灭火技能，尝试灭火

二、位置与语序

在实用翻译中，符合目标语行文习惯的译文，在词的方面，不仅要求词要译得准确到位，还要保证译文中每个词都放在正确的位置；在句子方面，不仅要求句子意思要准确理解，还要把握好原文句子隐含的侧重点，把句子的顺序合理化。

这里所讲的语序，主要包括each, every, some, any, all, part, other, another, relevant等小定语的位置，也包括其他定语的位置，还包括更复杂的括号内容（广义的词）位置的准确性。而对于扮演其他角色的句子成分的位置，可以依照类似办法予以确定。[①]

(1) 小定语位置

这里所讲的小定语，是指中文的"各个""每个""一些/有些""任何""全部/所有""部分""许多/很多""少量/一点点""其他""另一（个）""相关"等，及其相应的英文each, every, some, any, all, part, many, a little, other, another, relevant等，以及其他中英文短小定语。这些词虽然不是核心词，但也不得随意省译，除非译出来后显得很多余，尤其是在法律合同类文件，或者其他涉及严谨描述的文件里更是如此。而且，这些小定语的译文位置也必须尽可能地准确。例如：

例2.46：

原文：The Receiving Department is responsible for acceptance of shipments delivered in the Division's name, reporting any evidence of carrier-caused loss or damage, as described in Procedure G1401.

原译：接收部门有责任接收以事业部名义所交付的货物，同时公布任何由承运人所造成的损失或损坏的证据，如程序G1401所示。

改译：接收部门有责任接收以事业部名义所交付的货物，同时公布由承运人所造成的损失或损坏的任何证据，如程序G1401所示。

分析：原译把any evidence译成"任何……证据"，虽然符合语法，但因"任何"和"证据"的距离过远，显得句子不够紧凑。

例2.47：

原文：Other fees payable by the Company

原译：其他本公司应付费用

改译：本公司其他应付费用

[①] 对于由多个句子组成的句群，在翻译时，也必须根据语境，对句序加以调整。对句序的调整，请看后面的"句子顺序"部分。

分析：改译后，other的译文"其他"离other修饰的名词更近，这样不容易引起误解。

例2.48：

原文：Obtain all necessary permits and licenses required to operate your business.

原译：获得<u>所有</u>公司运营所需的许可证和执照。

改译：获得公司运营所需的<u>所有</u>许可证和执照。

分析：改译后，all的译文"所有"与其实际所修饰的词更接近。

（2）其他定语位置

例2.49：

原文：The review identifies two main security of supply challenges for the United Kingdom:

原译：该评论确定了<u>两项</u>英国供电安全挑战：

改译：该评论确认了英国供电安全的<u>两个</u>挑战：

分析：在本例中，句中的two指的是"challenges"的数量，这一点从challenges的复数形式可以看出。既然如此，就应该把two的译文"两个"尽量靠近"挑战"。

（3）括号位置

括号里的内容在句子中所扮演的角色，可以是定语、状语、同位语，也可以是补充说明。但是，不论是在中文还是英文中，不论括号内容在句式中扮演的是什么角色，其位置在原则上都必须尽可能靠近所修饰或说明的词。

例2.50：

原文：关于详细的工程计划，应与本公司制造部门协商，得到他们的施工允许。另外，本保温工事是本船建造上的重要工程，要遵守我公司所指示的各种（安全、质量、工期）事项。

原译：Specific engineering plans shall be subject to negotiation with and prior consent of the manufacturing department of our company. Moreover, the insulation works is critical to the building of the ship, all items <u>done</u> (relating to safety, quality and construction period) <u>by</u> our company shall be fully observed.

改译：Specific engineering plans shall be subject to negotiation with and prior consent of the manufacturing department of our company. Moreover, the insulation works is critical to the building of the ship, all items (relating to safety, quality and construction period) done by our company shall be fully observed.

分析：本例译文中done的位置，应该在括号之后，这是译文中词的位置的准确性要求。也就是说，如果括号里的内容是括号外内容的定语，则应把括号里的内容当作一个定语，摆在定语的位置。

例2.51：

原文：At around this time a number of emails were exchanged which, in so far as they expressed the direct views of the Company, show that it was taking an orthodox view of its contractual rights.

原译：大约在此期间，相关方之间往来的多封邮件（表达该公司的直接观点）表明，中国南方航空将按惯常行使其合约权力。

改译：大约在此期间，相关方之间往来的（表达该公司的直接观点）的多封邮件表明，该公司将按惯常行使其合约权力。

分析：本例对原文的理解和译文的表达都没错，但是括号的位置放得不好。

例2.52：

原文：When equipment is being transferred, scrapped or sold, all spare parts for that equipment (that will no longer be used at the site) should be identified and transferred with the equipment, scrapped or sold.

原译：设备转移、报废或出售时，确定该设备所有备件并依据设备处理情况转移、报废或出售备件。

改译：设备转移、报废或出售时，该设备的所有不再在原现场使用的备件应与该设备一起确认并转移、或报废或出售。

分析：本句由于句子不长，原文的括号，在译文中可以省去，但是必须把原文括号里的内容的译文放在准确位置。

（4）句子顺序

例2.53：

原文：All such items as pipes, fittings and valves, whether or not specifically mentioned or indicated there, shall be furnished and installed if necessary to complete the system in accordance with the best practices of the plumbing trade and to the satisfaction of the Employer's Representative.

原译：如果必要，管道、设备和阀门等所有项目，无论这里是否被提及或提示，都应被提供和安装，以完成符合水暖工作的最佳实践和雇主方代表的满意度的系统。

改译：为按照水暖行业的最佳做法，以业主代表满意的方式做好整个系统，如有必要，无论是否明确提及或指示，均应供应和安装管道、配件和阀门等。

分析：本例原译理解是准确的。但是，很明显的是，to complete后的内容是整个句子的目的，而且in accordance with the best practices of the plumbing trade和to the satisfaction of the Employer's Representative两部分，分别是complete的两个要求，都是complete的方式状语。因此，必须把if necessary和to complete后面所有内容的译文都移到译文的最前头，才是最符合原文本意的表达方式。

例2.54：

原文：The EPC Contractor shall submit for approval the necessary detailed drawings and complete list of materials to be incorporated in the Works, samples of materials with specimen of workmanship, colour and finish, manufacturers catalogues at least 60 days prior to the date he wishes to place or confirm orders.

原译：EPC工程总承包商应当把必要的详细图纸和工程所需材料的所有清单、材料样品（带工艺、颜色和饰面样本）、制造商目录，在下订单或确认订单之前60天提交，以供批准。

改译：EPC工程总承包商应当至少在下订单或确认订单之前60天，提交必要的详细图纸、工程所需材料的完整清单、材料样品（包括工艺、颜色和饰面样本）、制造商目录等，以供批准。

分析：原译意思表达并未出错，但整体句子显得断句不合理，不好理解，改译后调整了句子顺序，将时间状语提前，显得更容易理解。

三、目标语行文习惯

这里所强调的行文习惯，是指译文不仅要意思表达准确，而且行文方式要符合目标语的表达方式，不得出现过分的翻译腔或者句式生硬等现象。翻译腔或句式生硬的现象，有可能是由对原文的词或语法理解不够透彻所造成的，也有可能是由对目标语的遣词造句能力不够好所致。但是，无论哪种情况，都可以通过多思考、多练习加以改善。

（1）中文行文习惯

例2.55：

原文：All purchase requests are documented in sufficient detail and approved by the appropriate level of management before the purchase is made.

原译：所有采购申请都记录了详细细节，且在采购进行之前由适当的管理层进行批准。

改译：所有请购单都做了足够详细的记录，且在采购之前由适当的管理层批准。

分析：本例原译意思表达无误，但表"采购申请""记录了详细细节""进行批准"这样的表达不大符合中文行文习惯。

例2.56：

原文：The maximum desirable sustained gradient shall be 3.0%, however, where site conditions and/or geometric constraints dictate, the gradient may be increased up to but not exceeding 4.0%.

原译：理想的最大连续坡度应为3.0%，然而，受场地情况与/或几何限制的影响，可增加坡度但不得超过4.0%。

改译：理想的最大连续坡度应为3.0%，在受场地情况与/或几何限制的影响时，可适当增加坡度，但不得超过4.0%。

分析：原译文准确无误，但是句子不顺。对于这种情况，可以根据上下文，对译文进行适当的修改。

例2.57：

原文：The wind farms portfolio is currently completely under O&M contracts with the OEM's. This setup has been identified as solid and stable O&M strategy; however it is important to consider that for several wind farms, the O&M contract expires before the end of the lifetime of the wind turbines.

原译：本风电场投资组合当前完全包含在与原始设备制造商签订的运行和维护合同内。该计划已被稳固确立为运行维护策略，<u>但必须</u>考虑到多个风电场的运行维护合同在风力涡轮机使用年限达到前就已到期。

改译：本风电场投资组合当前完全包含在与原始设备制造商签订的运行维护合同内。该计划已被确认为稳定地运行维护策略，<u>但必须</u>考虑到<u>某些</u>风电场的运行维护合同<u>可能</u>在风力涡轮机使用年限达到前就已到期。

分析：在翻译中，译员可在准确表达原文意思的基础上，改变译文中的某些词的表达以使行文更通顺，如本句中将it is important to译成"但必须"就是很好的办法，改译后的译文将several从"多个"改成"某些"，并在句中加入"可能"一词，也是本着同样的原则。

例2.58：

原文：Proper implementation and management of equipment has a direct impact on process capability. This procedure is intended to outline the approach to equipment management. Specific details of how to implement the approach need to be developed and instituted through regional procedures and operating instructions.

原译：适当的设备运作和管理将直接影响流程性能。本程序旨在概述设备管理的方法。方法实施的具体细节需依据区域程序和操作说明制定。

改译：设备的运作和管理是否妥当，对工艺能力会产生直接影响。本程序将对设备管理方法做个大概的描述。而对于方法的具体实施，则需依据区域程序和操作说明予以制定。

分析：本例原译意思没错，但都翻译腔太重。总体上，译完后的每个句子，都要确保抛开原文来看译文时，感觉很通畅，很符合目标语（这里是指中文）的行文习惯。

例2.59：

原文：In preparation for acquiring equipment, funding must be approved prior to issuing purchase orders. <u>Prior to starting activity to acquire new equipment a significant effort should be made to locate equipment available in other</u>

Nexteer Automotive facilities that may be appropriate for the intended purpose. If appropriate equipment is available, it should be used as an alternative to purchasing new equipment.

原译：设备采购准备期间，必须获得资金批准方可发出订购单。开始采购新设备前，需花费大量精力确定其他耐世特汽车工厂内适用于预期目的的设备。如有适当设备，该等设备应作为替代，避免采购新设备。

改译：在准备采购设备期间，必须获得资金批准方可发出订购单。在开始采购新设备前，必须花大力气确定其他耐世特汽车工厂内是否有适用于预期目的的设备。如有适当设备，应将该设备作为替代，而非采购新设备。

分析：在加下画线部分的原译中，前一句意思表述不明，后一句表述不够顺。改译后，前一句意思更明确，后一句更符合中文行文习惯。

例2.60：

原文：Some lower kelly valve models are not designed to withstand external pressure encountered in stripping operations.

原译：一些下方钻杆阀模型并不是设计用于抵抗强行起下钻作业中所承受的外部压力的。

改译：一些下方钻杆阀模型的设计无法承受强行起下钻作业中所承受的外部压力。

分析：本句中的are not designed to withstand …的原译"并不是设计用于抵抗"，但如果就这样译出，是不符合中文行文习惯的，因此，要根据上下文意思改为"……的设计无法承受"。

例2.61：

原文：The transformer tank will be designed to withstand an overpressure of one bar as well as to allow near vacuum during drying-out before oil filling.

原译：变压器储能罐将设计成可承受1巴的过压，且在注油前的干燥过程中允许几乎真空。

改译：变压器油罐的设计应可承受1巴的过压，且在注油前的干燥过程中可承受接近真空的压力。

分析：原译句子明显不顺。不顺的原因，是译员对句子意思理解不透彻，因此句子只能跟着原文走。其实，像这种sth. be designed to withstand ... 的句式，通常是"某物的设计能承受……"之类的意思。而near vacuum则是指"几乎/接近真空"这样的压力。这里需说明的一点是，容器所承受的压力，即包括外部气压（通常是大气压）低于容器内部气压时产生的正向压差，也包括内部气压低于外部气压时所产生的反向压差，因此，这里将near vacuum表达成"接近真空的压力"。

例2.62：

原文：On average, the size of LCD displays has been growing one inch every year. Resolution has been increasing, and new product ideas — like curved displays — are entering the market. In a world of constant innovation, you need supreme quality to remain competitive. Your technology must also have a high yield in the LCD panel manufacturing stage of your production. Our Prexision systems are the industry standard for photomasks, used by all the world's photomask manufacturers.

原译：液晶显示器（LCD）的尺寸每年平均增长一英寸。分辨率一直在提高，新产品创意——例如可弯曲显示器——正在进入市场。在这样一个不断创新的世界上，您需要用至高的质量来保持竞争力。在液晶显示器屏幕的生产制造阶段，您的技术必须能够为您带来很高的产量。我们的Prexision系统是光掩膜的行业标准，全世界的光掩膜制造商都在使用我们的系统。

改译：近年来，液晶显示器（LCD）的平均消耗尺寸每年增长一英寸。而且，分辨率也一直在提高，新产品创意——例如曲面屏——也正在进入市场。在这样一个不断创新的世界上，您需要用至高的质量来保持竞争力。在液晶显示器屏幕的制造阶段，您的技术必须能够保证很高的产量。我们的Prexision系统是光掩膜的行业标准，全世界的光掩膜制造商都在使用我们的系统。

分析：

例2.63：

原文：Cement to be used for encasing of concrete shall be ordinary Portland and sand and coarse aggregate. Small lengths or cut-pieces of pipes with multiple jointing to make one single length under any road crossing shall not be permitted.

原译：包裹混凝土的水泥应为普通波特兰水泥及沙和粗骨料。在道路交叉口下方，不得使用多个短小或裁截管道连接而成的单段长度管道。

改译：包裹混凝土的水泥应为普通波特兰水泥及沙和粗骨料。在道路交叉口下方，不得以多个接头将较短的或切片的管道连接成一条管道。

分析：原译"使用多个短小或裁截管道连接而成的单段长度管道"表达得不顺畅，翻译痕迹明显，似乎理解不透彻，宜改成"以多个接头将较短的或切片的管道连接成一条管道"。

（2）英文行文习惯

例2.64：

原文：气压试验的系统不应过大，系统体积与试验压力的乘积不应超过50m³ MPa。

原译：The system of gas pressure test shall not be too large, and the product of system volume and testing pressure shall not exceed 50m³ MPa.

改译：The system of gas pressure test shall not be oversized, and the product of system volume and testing pressure shall not exceed 50m³ MPa.

分析：这里的"过大"是中式表达，不宜译成too large，译成oversized才是合理表达。

例2.65：

原文：当拧紧一组螺栓或螺母时，从中心或大直径螺栓开始，分两步或更多步以交叉方式来拧紧它们。

译文：When tightening a group of bolts or nuts, tighten them symmetrically from the center or large-diameter bolts in two or more steps.

分析：这里的交叉方式，实际是指对于呈圆环分布的螺栓和螺母，要对称地旋紧，否则会拧不紧或者出现损坏等问题点。

例2.66：

原文：景区外的配套设施为现代风格，预示着时代的发展，预示着和古代文化一脉相承的渊源。

本项目根据实际情况对穿越的交通干道进行一定的处理，使车行干道与步行干道相对独立，又紧密联系。

原译：Modern styles are used in supporting facilities outside the scenic spot, which indicates the development of the times, and the origin which comes down in one continuous line of the ancient culture.

According to the reality, crossing traffic arteries are processed in the project to make drive arteries and walking arteries relatively independent but closely connected.

改译：Outside the scenic spots are supporting facilities in modern style, implying the development of times and the heritage of ancient culture.

Traffic arteries crossing the project are treated based on practical conditions, to make driveway and walkway relatively independent but closely connected with each other.

分析：原译基本跟着原文的顺序走，显得很生硬，不符合英文表达法。改译后基本能符合英文行文习惯。

例2.67:

原文:以扎实的房地产与基础设施的专业知识为基础,<u>进军</u>地产PPP领域,为各级政府部门、社会资本企业提供PPP全程服务。

原译:Based on the rich professional knowledge in real estate and infrastructure, the team <u>marches towards</u> PPP area of real estate to provide one-stop services for governmental departments and social enterprises.

改译:The team <u>begins to engage in</u> PPP area of real estate based on its rich professional knowledge in real estate and infrastructure, to provide one-stop services for governmental authorities and social enterprises.

分析:注意,英语中的用词通常都是有固定搭配的。"向……进军"译成march toward是典型的中式英语译法,很多人还以为自己译得很好,却不知道英文里根本没这个表达。

例2.68:

原文:另外,由于本系统是基于Salve-Master的物理架构来<u>进行任务分配及调度</u>的,因此与目前市面上其他同类相比的话,<u>本系统还满足了多用户同时使用的需求</u>。

原译:In addition, the system <u>carries out task assignment and dispatch</u> on the basis of the physical structure of Salve-Master, therefore, compared with the similar systems available in the market, <u>the system also can meet the requirements that allow multiple users to use it simultaneously.</u>

改译:Moreover, the system <u>assign and deploy tasks</u> based on the Salve-Master physical structure, therefore, compared with similar systems in current market, the system also allows multiple users to use it simultaneously.

分析:本句加下画线的部分,原译把原文出现的字基本一字不落地译出来。但是,"carry out + 动词的名词形式",在英文里使用频率并不是很高,在本句中这样使用是不妥的。另外can meet the requirements that allow multiple users to use it simultaneously的句式,总体上也偏中式。而改译后则总体上比较符合英文表达习惯。

例2.69:

原文:观察10分钟,压降不超过100psi,<u>合格</u>。

原译：Be qualified if the pressure drop is less than 100psi after observed for 10 minutes.

改译：It is OK if the pressure drop observed is less than 100psi within 10min.

分析：be qualified通常用于表示人"称职""有资格"，较少用于表达产品或性能符合要求。本句可根据上下文将"合格"译成OK。

例2.70：

原文：打开高压截止阀放压，放压为0MPa，关闭高压截止阀。

原译：Open the high-pressure check valve to release pressure. If 0MPa is released, close the high-pressure check valve.

改译：Open the high-pressure check valve to release pressure to 0MPa, and then close the high-pressure check valve.

分析：这里的"放压为0MPa"，指的是将压力完全卸掉，卸到接近无压力（压力为0MPa）。这对于有理工背景的译员来说，不会很难判断，但没有理工背景的译员却很容易误入歧途，导致字面翻译，从而让内行者感觉很异常。

所以，这里介绍一点相关知识。平常我们所处的空间，充满着空气，空气是有压力（压强）的，其压力是一个大气压（1atm），1atm = 101325Pa≈0.1MPa。而油气钻井中的高压，通常是以MPa（兆帕）为单位的，如果压力接近0MPa，就是压力接近于零了。

本节课后练习

翻译下列句子，注意下画线部分：

1. 所有需从合作方进口的精密设备均应由专业人员安装。

2. Any deviation from these standards must be approved in writing by the Montague Operations hydraulic engineer.

   …

   The final Bill of Material shall list recommended spare parts list.

   （请注意下画线词的位置）

3. For the extraction of juice from citrus fruits (oranges, lemons, mandarins, grapefruits, limes).

4. most important symptoms and effects, both acute and delayed
5. The pressure relief valve protects the system from excessive pressure. It is factory set.
6. An operator's cabin shall be provided in the crane.
7. Economic growth in developing countries is especially important because food consumption and feed use are particularly responsive to income growth in those countries, with movement away from staple foods and increased diversification of diets.
8. The modernized railway line shall enable commercial speed of 130 km/h for the fastest passenger trains and reduce Belgrade-Budapest travelling time to less than 3.00 hours. Besides speed, a modern double track railway line shall ensure high safety, capacity and comfort for passengers and goods. That will be a significant contribution to full capability of the railways compared with other modes of transport and will enable rational redistribution of traffic also enhancing the level of environmental protection.
9. 所采用电线应具有耐候性，适合热带气候条件，并具有可承受1,100V电压的聚氯乙烯绝缘层。
10. 可把电吹风对准起褶部位来回加热，当门封条变软时，将其调整后待自然冷却后定型即可。
11. 当动物同时接触两条相线或同时接触相线和大地时，就会触电。
12. 洋葱可分为普通洋葱、分蘖洋葱和顶生洋葱。
13. 有些研究者利用在人工生态浮岛上进行营养液挂膜吸附、截留水中污染物的技术来净化水体。

第三章　中文理解与表达

第一节　翻译相关中文基础知识

一、中文连词搭配

在英译中时，不仅要将中文意思表达出来，还要注意中文用词，包括中文连词的使用是否准确等。以下是很多译员不习惯使用的几组中文连词搭配：

除非……否则、如果/如/若/一旦……则/那么、尽管……仍、所有/任何……均/都、只要……就、虽然……但是、因为……所以

以上连词搭配，基本上是按照译员出错频率由高到低排列的。以下是前四组连词搭配的例子：

（1）除非……否则

例3.1：

原文：Do not use compressed air for breathing purposes unless it is specifically treated.

Do not use compressed air for any application that will bring it into direct contact with foodstuffs unless it is specifically treated.

原译：<u>除非经过特殊处理</u>，请勿将压缩空气用于呼吸目的。

<u>除非经过特殊处理</u>，请勿将压缩空气用于可能与食物产生直接接触的用途中。

改译：<u>除非经过特殊处理</u>，<u>否则不得</u>将压缩空气用于呼吸目的。

<u>除非经过特殊处理</u>，<u>否则不得</u>将压缩空气用于可能与食物产生直接接触的用途中。

分析：原译句子意思没错，但是，中文的"除非"一般是要与"否则"相互搭配的。

（2）如果/如若/一旦……则/那么

例3.2：

原文：Grouting will be regarded as being satisfactory if the pressure can be maintained for at least 5 minutes without further grout take.

原译：在无进一步吃浆量的情况下，如果压力能够维持至少5分钟，认为灌浆满足要求。

改译：在无进一步吃浆量的情况下，如果压力能够维持至少5分钟，则认为灌浆满足要求。

分析："如果/如若/一旦……则/那么"是汉语中比较正式的连词搭配，其中"那么"比"则"更口语化，"如果"比"如若"更口语化。

（3）尽管……仍

例3.3：

原文：Although the spherical tank meets the user fatigue operating conditions, frequent and significant pressure fluctuations in operation is not conducive to the fatigue strength and should be avoided if possible. The operating pressure should be kept steady.

原译：尽管本球罐满足用户的疲劳操作条件，操作中频繁的大幅度压力波动，对储罐的抗疲劳强度是不利的，应尽可能避免，保持操作压力平稳；

改译：尽管本球罐满足用户的疲劳操作条件，操作中频繁的大幅度压力波动，对储罐的抗疲劳强度仍是不利的，应尽可能避免，保持操作压力平稳；

分析："尽管/固然/纵然……仍/然而"是常见的汉语连词搭配，有时"但是""仍"会在"尽管/固然/纵然"后同时使用。

（4）所有/任何……均/都

例3.4：

原文：PVC cement glue is coated by HP spray gun, with all working parameters controlled by the HM interface and the demonstrator in a closed loop.

译文：PVC胶采用高压喷枪喷涂，其所有工作参数均通过人机界面及示教器进行闭环控制。

分析："所有/任何……均/都/概/一概"也是常见的汉语连词搭配。

二、易用错的中文词

汉语中有些词相互之间字形、读音或意思很相近，很容易被写错和/或理解错。其中有些词在实用翻译里出现的频率还比较高，而且用错了之后也不易发现。以下所列基本都是因为形、音、义近似而容易写错的词，有些词同时还很容易理解错。

（1）阈值—阀值

例3.5：

原文：Emergency-Depressurisation (EDP): Control actions undertaken to depressurise equipment or process down to a pre-defined threshold (generally 7 barg or 50% of design pressure) in a given period of time (generally 15 minutes) in response to a hazardous situation.

译文：紧急降压：系指在响应危险情况时，在特定时间段（通常为15分钟）内，使设备或工艺压力降低至预定阈值（阀值）（一般为7 barg或者设计压力的50%）所采取的控制措施。

分析：threshold的中文意思的正确写法是"阈值"，而不是"阀值"。"阈值"和"阀值"是易混淆的两个不同概念，而且人们普遍会将"阀"的意思与"阀门"相关联，所以更容易认为"阀值"是threshold的正确译文。但实际上，"阀值"只是"阈值"的笔误。

（2）抑或—亦或

例3.6：

原文：Foreign buyers would be interested mainly in access to Malaysian market, not in Proton's factories, models or headstrong managers, who insist that a little more investment is all that is needed to turn the firm around.

译文：国外买家感兴趣的主要是能进入马来西亚市场，而不是宝腾的工厂、车型抑或（亦或）刚愎自用的管理者。这些管理者坚持认为要扭转公司颓势只需再增加些许投资即可。

分析："抑或"的意思是"还是""或是"，而"亦或"则是不规范或者错误的汉语表达。

（3）泄漏/泄露

例3.7：

原文：Containers that have been opened must be carefully resealed and kept upright to prevent leakage.

译文：已打开的容器应妥当地重新封闭，并保持直立，以防出现泄漏（泄露）。

分析："泄漏"和"泄露"两个词发音相同，但两者意思有所区别。两个词的意思分别为：

泄漏：①让人知道不该知道的事；②（气体或液体）渗透、漏出。

泄露：①让人知道不该知道的事；②显露、显现；③暴露。

也就是说，在透露消息方面，"泄漏"包含有"泄露"的意思。

（4）体系/系统

例3.8：

原文：We would expect implementation of ISO 28000:2007 which specifies the requirements for a security management system, including those aspects critical to security assurance of the supply chain.

译文：我们希望能够执行ISO 28000:2007对安全管理体系（系统）做出的要求，其中包括对供应链安全保证起到关键作用的要求。

分析：注意，在程序管理方面，system一般译为"体系"，不译成"系统"。"体系"和"系统"的区别如下：

体系

从词义上讲，体系（system）是一个科学术语，泛指一定范围内或同类的事物按照一定的秩序和内部联系组合而成的整体。如质量管理体系、环境管理体系、职业健康安全管理体系、（热力学）敞开体系/封闭体系/孤立体系。

系统[①]

解释一：同类事物按一定的关系组成的整体。如组织系统、灌溉系统、计算机系统、教育系统。

解释二：有条有理的。如系统学习、系统研究。

（5）于/予/与

例3.9：

原文：Preventative and proactive approaches should be developed and adopted to empower vulnerable people and effectively integrating them into the wider community.

译文：应制定并采用预防性的积极措施，授予（授于）弱势群体权利，以有效地使弱势群体融入更大的社会团体中。

分析："授予"的意思是给予荣誉、学位、权利等，而"授于"则是不规范用法。由电脑输入法《在线汉语字典》与《新华字典》均可判断用词是否规范。

小结：

以上例子只是翻译中经常碰到的一小部分易写错且易理解错的词，由于这类词数量很多，现将翻译中经常碰到的这类词列表如下：

易错词	说明
阈值/阀值	"阀值"是"阈值"的笔误
抑或/亦或	"亦或"是"抑或"的笔误
泄漏/泄露	泄漏：①让人知道不该知道的事；②（气体或液体）渗透、漏出。 泄露：①让人知道不该知道的事；②显露、显现；③暴露。
试验/实验	实验：为了检验某种科学理论或假设而进行某种操作或从事某种活动。 试验：为了察看某事的结果或某物的性能而从事某种活动。
体系/系统	体系：科学术语，泛指一定范围内或同类的事物按照一定的秩序和内部联系组合而成的整体。如质量管理体系、环境管理体系。 系统：同类事物按一定的关系组成的整体。如组织系统、灌溉系统、计算机系统、教育系统。

---

① 此解释源自《中华大词典》。汉语大字典编纂处编著，《中华大词典》，四川：四川辞书出版社，2020。

（续表）

易错词	说明
于/予/与	这三个字都可放在别的动词后面，组成一个较为固定的词语，表示"给"的意思。如归于、授予、赠与等。一般情况下，"于"和"与"的对象多为人，如荣誉归于大家、信件已交与本人。"予"的对象一般指物，如授予勋章等。 "予"和"与"可以单独使用表示"给"的意思，如予以处理、与人方便等；而"于"没有这种用法。 "于"可以放在复合动词（两个字组成）后面表示"给"的意思，如献身于科学事业。而"予"和"与"则没有这种用法。
以至/以致	"以至"有两种意思，其一是"直到""一直到"。如"熟练的技能是经十次、百次以至上千次的练习才能获得的。" "以至"的另一意思是达到某种程度或结果。如"科学发展日新月异，以至于神话、童话故事里的幻想也变成了现实"。 "以致"为连词，用在下半句开头，表示下文是上述情况造成的结果，多指不良后果。如"他摔伤了，以致几个月不能起床"。
雇用/雇佣	"雇佣"只作定语，如雇佣兵、雇佣观点、雇佣劳动（见《现代汉语词典》）和雇佣合同（见《辞海》）。 "雇用"是一般及物动词，如雇用临时工、雇用合同工。
蒸汽/蒸气	"蒸汽"特指水加热到沸点而生成的水汽，义同"水蒸气"，所以"水蒸汽"的写法不对。 "蒸气"泛指液体或固体（如水、碘、汞、苯）因蒸发、沸腾或升华而生成的气体。如水蒸气、苯蒸气、汞蒸气、碘蒸气。
电气/电器	电器泛指所有用电的器具，通常指家庭常用的为生活提供帮助的用电设备。 电气是指电能的生产、传输、分配、使用和电工装备制造等学科或工程领域的统称。
浇注/浇铸/浇筑	强调浇灌动作时，用"浇注"；强调成型时，冶金行业用"浇铸"；水泥等其他行业的浇灌则用"浇筑"。
收集/搜集	"收集"侧重于"收"，即收拢，对象是已有的事物；"搜集"则重于"搜"，即搜寻，对象是需花时间和精力寻找的事物。
截止/截至	截止：到某时间为止，强调"止"，后面不跟时间做宾语。如"报名于今天截止"。 截至：停止于某个时间，强调时间，后面要跟时间做宾语。如截至目前。

（续表）

易错词	说明
定金/订金	根据我国《中华人民共和国民法通则》和《担保法》规定，定金与订金的区别主要表现在四个方面： 1. 交付定金的协议是从合同，依约定应交付定金而未付的，不构成对主合同的违反；而交付订金的协议是主合同的一部分，依约定应交付订金而未交付的，即构成对主合同的违反。 2. 交付和收受订金的当事人一方不履行合同债务时，不发生丧失或者双倍返还预付款的后果，订金仅可作损害赔偿金。 3. 定金的数额在法律规定上有一定限制，例如《担保法》就规定定金数额不超过主合同标的额的20%；而订金的数额依当事人之间自由约定，法律一般不作限制。 4. 定金具有担保性质，而订金只是单方行为，不具有明显的担保性质。 由此可见定金和订金虽只一字之差，但其所产生的法律后果是不一样的，订金不能产生定金所有的四种法律效果，更不能适用定金罚则。
国外/外国	"国外"是指除了本国政治领属之外的地方，常用于指本国以外的具体或抽象事务或概念。"外国"是指除了本国之外的国家或地区，范围比较小。这两个词的习惯用法有所不同，通常说"国外产品"，不说"外国产品"，说"外国人"不说"国外人"。

三、易误译的中文词

汉语中有些词因存在一词多义现象而让人误解，译者按自己错误的理解翻译，有些甚至会出现与原文截然不同或者截然相反的错误译法。以下所举例子是比较常见的误译词。

（1）开/关

例3.10：

原文：打开开关时，灯就会亮。

原译：A bulb will light when its switch is open.

改译：A bulb will light when its switch is turned on.

分析：本例是非常典型的例子，原译和改译意思完全译反。
中文的"打开开关"，通常是指让电路通电。但是英文的the switch is open却相当于the switch is turned off，即电路断开、不通电的意思。

因为电的开关,很多都是闸式的,合上(close)闸式开关时,就是通电,打开(open)闸式开关时,就是断电。

同理,open circuit是"断路""开路",意思是电路不通电,close circuit是"闭路",意思是电路通电。所以,为保险起见,建议将"打开开关"译成turn on the switch,将"合上(闭合)开关"译成turn off the switch,以免误解。如果是英译中,open the switch最好译成"断开开关",close the switch最好译成"闭合开关"。

(2)高压/低压

例3.11:

原文:当物质在高压下被压缩时,其物理、化学性质会发生改变。

译文:Physical and chemical properties of a matter will be changed when it is compressed under high pressure.

分析:本例的"高压"是指环境压力,只能译成high pressure。

例3.12:

原文:高压绝缘手套是电工安防用品,可对手或者人体起到保护作用。

译文:High voltage insulation gloves are electric safety products protecting the hands or human body.

分析:本例的"高压"是指电压,只能译成high voltage。

例3.13:

原文:他今天高压150mmHg,低压80mmHg。

译文:His systolic pressure is 150mmHg and diastolic pressure 80mmHg today.

分析:本例的"高压"和"低压"分别指人体血管的收缩压与舒张压,一般译成systolic pressure和diastolic pressure。

以上三例的"高压/低压",中文意思各不相同,因此译法也各不相同,把任何一个"高压"或"低压"译成其他译法,都是错误的。

(3)指数

例3.14:

原文:中国奢侈品消费以指数倍增长。

译文:The consumption of luxury goods in China witnessed an exponential growth.

分析：本例的"指数"是一个源自数学的经济学概念。在乘方a^n中，a为底数，n为指数，结果为幂。当一个变量从一个时期以固定比率增长时，就叫作指数倍增长，也叫几何增长或几何级数增长。

例3.15：

原文：以下<u>采购经理人指数</u>由中国国家统计局发布。

译文：The following <u>purchasing managers index</u> (PMI) data was updated by the National Bureau of Statistics of China.

分析：本例的"指数"也是一个常用的经济学概念。广义地讲，任何两个数值对比形成的相对数都可以称为指数。狭义地讲，指数是用于测定多个项目在不同场合下综合变动的一种相对数。常见的经济学指数包括消费者价格指数CPI、采购经理人指数PMI、工业价格IPPI以及各大证券交易所的股票指数、证券指数等。

（4）折扣

例3.16：

原文：凡家长携孩子进店参与活动即可享受本店产品购物低至六折的<u>折扣优惠</u>。

译文：All parents taking the kids joining in the activity may enjoy a <u>discount</u> up to 40% when buying goods in our shop.

分析：本例的"折扣"，是指降价销售，几折就是百分之几十，如六折就是百分之六十的价格。

例3.17：

原文：公司政策的执行效果被<u>打折扣</u>了。

译文：The implementation effect of the company's policy <u>is weakened</u>.

分析：本例的"打折扣"，是指效果不好。

例3.18：

原文：他的承诺必须<u>打折扣</u>。

译文：His commitments is unbelievable.

分析：本例的"打折扣"是人或话不可信的意思。

（5）随机

例3.19：

原文：采样方案必须确保只从地下矿区和露天矿区随机采集大块样品。

译文：The sampling plan must ensure that bulky samples are only randomly picked from underground or open mining areas.

分析：本例的"随机"指的是"随意""毫无规律"的意思。

例3.20：

原文：本产品随机提供相关说明书、图纸及少量易损备件。

译文：The product is attached with relevant specifications, drawings and some easily worn spare parts.

分析：本例的"随机"不是指"随意"，而是指"随附于机器"的意思。这种表达在设备或机械的说明书非常常见。

此外，"随机"还有"根据情势"的意思，如"随机应变"。

（6）改变/更改/变更/更换

例3.21：

原文：承包商提出的设计变更以及对材料、设备的更换，须经工程师同意。

原译：The design modification and change of materials and equipment proposed by the Contractor shall be agreed by the Engineer.

改译：The design change and replacement of materials and equipment proposed by the Contractor shall be agreed by the Engineer.

分析：本句中的"设计变更"，是指设计发生变化（change），但不一定是改进（modification），因此，译成design change比design modification好。"材料、设备的更换"是指将一种材料或设备替换成另一种，强调的是替换（replace），而不是变动（change），因此用replacement，不用change。

（7）自由/自主

例3.22：

原文：在牛顿力学中，自由落体运动是指物体只受到重力作用时所发生的运动。

译文：In Newtonian physics, free fall is any motion of a body where gravity is the only force acting upon it.

分析："自由"当形容词用，表示不受限制和约束时，通常可译成free。

例3.23：

原文：可燃冰的成功开采，标志着我国实现了天然气水合物勘查开发理论、技术、工程、装备的完全自主创新。

译文：The success in production of flammable ice marks the independent innovations by China in theory, technology, engineering and equipment for natural gas hydrate exploration and exploitation.

分析："自主创新"的"自主"，是指独立、不受他人支配，因此译成independent innovation，而不是free innovation。这是初学者很容易犯的错误。同理，公司的"自主经营"应译成independent operation。

（8）问题

例3.24：

原文：朝鲜核问题，是指朝鲜开发核应用能力而引起的地区安全和外交等一系列问题。

译文：North Korea's nuclear issue is a series of regional safety and diplomatic issues caused by the development of nuclear application capability of North Korea.

分析：本例的"问题"，是指政治上的大事或议题，应译成issue。

例3.25：

原文：这个问题很难回答。

译文：This question is hard to answer.

分析：本例的"问题"是指疑惑、疑问，是需要回答（answer）的。学习中遇到的"问题"通常都译成question。

例3.26：

原文：环境污染会导致一系列健康问题。

译文：Environmental pollution can result in many health problems.

分析：本例的"问题"是指客观存在的或遇到的疑难问题，有时很严重，是需要加以解决（solve，work out）的。

例3.27：

原文：一般而言，公司禁止上班时间处理私人问题。

译文：Generally, the company does not allow to deal with private matters in working time.

分析：本例的"问题"是指需要花时间应对的事情、麻烦事等。

例3.28：

原文：这台机器出问题了，得修一下。

译文：The machine is in trouble, it must be repaired.

分析：本例的"问题"是指"故障""异常情况"等。

（9）处理/处置

例3.29：

原文：城市污水的处理程度及处理工艺一般根据其利用或排放点确定，并考虑水体的自然净化能力。

译文：The treatment degree and process of municipal sewage is generally determined by its usages and discharge points while considering the natural purification capability of water body.

分析：本例的"处理"，是指对废弃物等通过特定工艺手段予以无害化，一般译成treat/ treatment。此外，在英文思维里，治病是对疾病进行"处理"，因此使用treat的名词形式treatment表示"治病"。

例3.30：

原文：平板玻璃制造、处理和加工用机械和设备——安全要求——第4部分：倾斜台。

译文：Machines and plants for the manufacture, treatment and processing of flat glass — Safety requirements — Part 4 : tilting tables.

分析：本例的"处理"，是指以一定的工艺程序对产品进行优化。类似的还有金属的"热处理"（heat treatment），对设备或工件的"表面处理"（surface treatment）以及"预处理"（pre-treatment）等。

例3.31：

原文：可编程信号的处理。

译文：Processing of programmable signals.

分析：信号等数据的"处理"一般译成process。相应地，"中央处理器"译为central processing unit（CPU），而一般的"处理器"则可译成processor。

例3.32：

原文：废弃机器的处理/处置收益归公司所有。

译文：The disposal proceeds of waste machines belong to the company.

分析：本例中的"处理"，是指对废弃物进行销毁、转让、变卖等处置，意思与"处置"相同，故译成disposal，其动词形式为dispose of。

例3.33：

原文：本协议项下所有争议均应由双方指定的仲裁机构处理。

译文：All disputes under the Agreement shall be settled by the arbitration agency designated by both parties.

分析：争议的"处理"是指对争议的"解决"，故译成settle。

例3.34：

原文：人事部应恰当地处理各种复杂事务。

译文：The personnel department shall properly deal with all complicated matters.

分析：这里的"处理"，是指做好有一定难度的事情，一般用deal with，如果是"处理"难度极大的事情，可以译成cope with。但这样的中文表达总体上不是很正式。

例3.35：

原文：这种机器的金属型材所有表面都需进行去毛刺处理/热镀锌防腐处理。

译文：All surfaces of the metal profiles of this machine must be deburred/galvanized for anti-corrosion.

分析：本例的两个"处理"都是无意义成分。其中，"进行去毛刺处理"就是"去毛刺"，"进行热镀锌防腐处理"实际上就是"为防腐而热镀锌"，因此"进行"和"处理"均不译。

（10）一次/一次性

例3.36：

原文：变压比是二次电压与一次电压之比，在忽略漏磁的情况下，等于二次线圈与一次线圈的匝数之比。

译文：Transformation ratio is the ratio of secondary voltage to the primary voltage and equal to the ratio of the number of turns in the secondary winding to the number of turns in the primary winding, if leakage flux is neglected.

分析：一次线圈、一次电压与二次线圈、二次电压分别是变压器一次侧与二次侧所用的线圈及线圈上的电压。其中"一次"也可称为"初次""初级"，"二次"也可称为"次级""二级"，"一次线圈"也称为"原线圈"，"二次线圈"也称为"副线圈"。一次、二次分别译为primary和secondary。

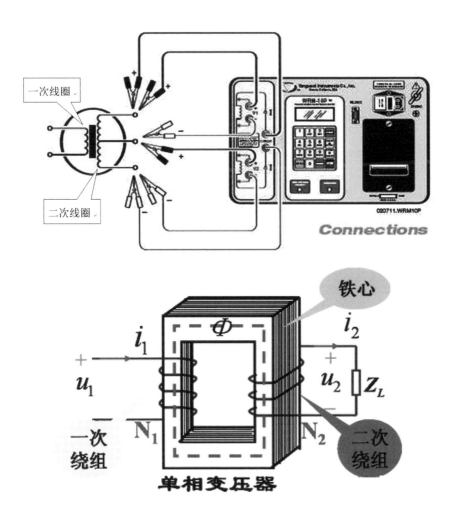

例3.37：

原文：当正弦电压被接到非线性负载上，基波（一次谐波）电流发生畸变时，即产生二次谐波、三次谐波和更高次谐波。

译文：Second harmonic, third harmonic and higher-order harmonics are generated when fundamental wave (first harmonic) current is distorted after sinusoidal voltage is connected to a nonlinear load.

分析：本例中的"一次谐波""二次谐波""三次谐波"和"更高次谐波"分别译为first harmonic、second harmonic、third harmonic、higher-order harmonic。"一次谐波"（first harmonic）就是电流"基波"（fundamental wave或original wave），而二次谐波、三次谐波和更高次谐波则是电流中所含有的频率为基波的整数倍的分量。

例3.38：

原文：本政策中所载明的旅行费用将仅依据提交的收据进行报销，但有关对食物和住宿采取一次性付款补贴的情形除外。

译文：Travel expenses as stated in this policy can only be reimbursed subject to the submission of receipts, with the exception of lump sum payment for food and accommodation allowance.

分析：本例的"一次性付款补贴"，是指补贴一次性发放。在英文思维里，"一次性"的付款被理解成整体性付款，因此译成lump sum（总额，全部金额）。类似的用法还有lump sum contract（总价合同，包干合同）。

例3.39：

原文：经过质量小组的共同努力，涂装车间一次下线合格率首次突破85%。

译文：With the joint efforts of our quality team, the first time quality (FTQ) of Paint Shop exceeded 85% for the first time.

分析："一次下线合格率"是指产品首次经过生产线时的合格率，在业内译成first time quality (FTQ)或first time pass rate (FTP)。

例3.40：

原文：如果必须触碰客人的食物，应使用勺子和叉子或者戴上一次性的塑料手套。

译文：If guest's food must be handled, use a spoon and fork or wear disposable plastic gloves.

分析：这里的"一次性"是指"用后即弃"，disposable本意是"可丢弃的"，引申意为"一次性的"，而disposables则是"一次性用品"的意思。

例3.41：

原文：在计算拟投资项目净现值时，应将项目前期市场调查费用计入一次投资成本。

译文：When calculating the net present value of proposed investment project, the preliminary market investigation fee should be included into the initial investment cost.

分析："一次投资成本"是"初始投资成本"的不规范表达法。

本节课后练习

请说明以下各组词的区别：

阈值—阀值
抑或—亦或
泄漏—泄露
体系—系统
于—予—与
雇用—雇佣
蒸汽—蒸气
电气—电器
浇注—浇铸—浇筑
收集—搜集
截止—截至
定金—订金

第二节　常见中国特色表达

汉语与英语的差异非常大，这些差异既体现在词语所蕴含的意义上，有时也体现在词语的文化内涵上。本节介绍的中国特色表达，在中国人看来多数都是很常见的表达。但真正要翻译时，有时我们竟然会发现，我们其实并没有真正理解这些词的意思。所以，本节介绍的中国特色表达，真要地道地译成英文，还是要经过一定的思考和转化的。这些词语主要包括动词、名词、形容词等，也有少量虚词。具体请看以下内容。

一、中国特色动词

中文有一类动词，如："进行""予以""经过""处理""采用/选用""结合"等。这些词有些本身并不是具体的动作，还必须与其他词配合使用才能表示具体的动作。另外，还有一些动词后面加了"好"或"做好后"等字。

对于这些词的翻译，很多新译员都会本能地按字面直译成carry out，go through，do，do well，process/ treat，combine，do good等。殊不知，这就陷入了过分直译，甚至逐字翻译的误区，这种译文不仅啰唆，还会让非汉语母语读者感觉怪异、不自然，甚至影响句子的理解。

其实，对这些词的正确译法，是要弄清楚这些词所表示的真正动作，即真正想表达的意思，然后尽可能用这些词所表示的真正动作作为句子的动词，译出真正意思。当然，如果直接以"进行""予以""经过""处理""好""结束（完成）后""结合"等所对应的英文译出，能使句子很顺、很自然，而且符合英文表达法，那也可以用这样的直译法。以下是对此类动词的译法分析：

（1）进行/实施、予以、进行/经过/予以……处理

例3.42：

原文：乙方主要工作之一是对项目组需求调研人员<u>进行</u>方法论培训。

原译：One major task of Party B is <u>carrying out</u> methodology training for demand researchers of the project team.

改译：One major task of Party B is <u>providing</u> methodology training for demand researchers of the project team.

分析：本句的"进行方法论培训"，其实就是"给予/提供方法论培训"。

例3.43：

原文：解释人员应根据相关规范，使用相应解释软件对测井资料进行数据处理。

原译：The interpreters shall carry out processing for the logging data with corresponding interpretation software in accordance with relevant specification.

改译：The interpreters shall process the logging data with corresponding interpretation software in accordance with relevant specification.

分析："对测井资料进行数据处理"的意思，就是"处理测井数据"，本句中"资料"和"数据"意思基本相同。其中"进行"常用的中文辅助动词，无实质意义。

例3.44：

原文：端头/边缘应按照以下规定方法进行刨平和切割或保持轧制状态。

原译：End/edge planning and cutting shall be done by anyone of the methods prescribed below or left as rolled.

改译：Ends/edges shall be planed and cut or left as rolled by anyone of the methods prescribed below.

分析：本句原译和改译都是挺顺畅的表达，但后者还是更直接更简洁明了。原译将"进行"译成be done，感觉上也比be carried out更简单一些。

例3.45：

原文：支吊架系统固定件、吊杆、横担（含螺栓、螺母）等配件表面需经过镀锌处理。

原译：The surfaces of such parts as fasteners, suspenders, cross arms (including bolts, nuts) of the support and hanger system shall go through galvanization treatment.

改译：The surfaces of such parts as fasteners, suspenders, cross arms (including bolts, nuts) of the support and hanger system shall be galvanized.

分析：本句的"经过镀锌处理"的意思，其实就是"镀锌"，因此只需译成be galvanized。其中常用的中文辅助动词"经过"，无实质意义。

例3.46：

原文：应将实施熏蒸之日起三天后的天气，进行准确预测。

原译：For the three days after carrying out the fumigation, the weather should be correctly predicted.

改译：The weather of the following three days after fumigation shall be correctly predicted.

分析：本例也是典型的中国特色表达。原译没有把"进行"译成carry out，但把"实施……"译成了carry out。其实，中文表达"实施熏蒸之日起三天后"的意思就是"熏蒸之后三日"，而"应将……进行预测"的意思就是"应预测……"

例3.47：

原文：对事故外逸的有毒物质和可能对人和环境继续造成危害的毒物，应及时组织人员予以消除。

译文：Personnel shall be organized to eliminate the poisonous matters that released in the accidents and that will continue do harm to human and environment.

分析：本例的"予以消除"，其实就是"消除"的意思，因此直接译成eliminate。

引申说明：

Carry out/conduct/do + -ing或carry out/conduct/do + 名词化动词的结构，或者其被动式，即V-ing be carried out/conducted/done，以及go through + -ing或go through + 名词化动词的结构，在英文里使用频率并不高，尤其是carry out的表达。所以，在英译时，能够不用这种句式的，就尽量不用，即使要将其译出，也可以carry out、conduct、do等几个词变换使用。

同理，在英译中时，尽管国内读者理解能力比较强，但也应当适当减少"进行""经过"之类的词的使用。如：

例3.48：
原文：The Contractor shall maintain a continuous record of the volume of concrete used and the level of the concrete in the pile.
原译：对于所用混凝土量以及桩身内混凝土等级，承包商应进行连续记录。
改译：承包商应连续记录所用混凝土量以及桩身内混凝土等级。
分析：（1）如果英文句子比较长而复杂，不好表达，有时可以使用中文的"对于"句式。但尽量少用，因为这种句式虽然没错，但不直接，不干脆，不符合应用类文件简洁扼要的要求。
（2）中文里不要有太多的"进行"或"予以"加动词的表达法，而是直接使用实际动词，才能使句式更简单。

（2）好、完成（结束）后

例3.49：
原文：中标人应基于国际最佳实践和本公司工程技术现状，<u>做好技术差距分析</u>。
原译：The bid winner shall <u>do well technical gap analysis</u> based on international best practice and the status quo of the Company's engineering technology.
改译：The bid winner shall <u>complete technical gap analysis</u> based on international best practice and the status quo of the Company's engineering technology.
分析："做好技术差距分析"的意思，实际上就是"分析技术差距"或"做技术差距分析"，"好"的意思是"完成"，不是"差""坏"的反义词，可以译成complete（或make）technical gap analysis，但不能译成do well technical gap analysis。

例3.50：
原文：灌浆时，必须<u>保护好</u>原钢衬结构的防腐等保护层。
原译：During grouting, the protective layers such as the anti-corrosion coating of the original steel liner structure shall be <u>well protected</u>.
改译：During grouting, the protective layers such as the anti-corrosion coating of the original steel liner structure shall be <u>protected</u>.
分析："保护好"里的"好"的意思是"完成"，"保护好"就是"完成保护"，就是"保护"，而不是"保护得不错"。

例3.51：

原文：钢衬接触灌浆结束后，清理螺纹孔，并做好后续的修补工作。

原译：After completion of contact grouting of steel liner, clean up thread holes and do well subsequent repair work.

改译：Clean up thread holes after contact grouting of steel liner, and complete subsequent repair work.

分析：在将"……完成后""……结束后"译成英文时，通常不必译成after the completion of，只需译成after ...（动名词或名词化动词）即可。另外，"做好"的意思就是"完成"，所以译成complete。

例3.52：

原文：解析完成之后，针对无须进行保管的不良品，虽然已有实施消毒对应，为保险起见，使用部门在对解析品完成解析后，予以实施蒸汽灭菌处理后，作为一般垃圾进行处理。

原译：After the analysis is completed, for defective products that need not to be preserved, even though disinfection treatment has been conducted, for insurance purposes, analyzed products shall go through steam sterilization after the analysis is done by the user department, and be handled as general waste.

改译：After the analysis, though defective products that need no preservation have been disinfected, to avoid risks, the user department must dispose of them as general wastes after the analysis and steam sterilization of them.

分析：本句是很典型的中国特色表达，有点口语化。原译总体上是跟着原文的句式翻译，是很典型的中式英文。

其实，原文的意思，可转化为：

在解析后，对于无须进行保管的不良品，虽然已经消毒过，但为保险起见，使用部门应将不良品解析和蒸汽灭菌后，作为一般垃圾处置。

其中，原文加下画线的部分，建议按改译中体现的方式翻译，也就是说，不能按字面直译成英文，而是应该先将其转化为更简洁、更有逻辑性的表达，发现原文的真正动词，再直译成英文。

（3）采用/选用

例3.53：

原文：此段内表面应<u>采用</u>平镀锌板，可耐高温并且便于清洁。

原译：The surface of this section shall <u>adopt</u> flat galvanized plate that is easy to clean and with high temperature resistance.

改译：The surface of this section shall <u>be made of</u> flat galvanized plate that is easy to clean and resistant to high temperature.

分析：本句的"采用"是指"由……制成"，要译成 be made of/from，不得出现 adopt/use 之类的词。

例3.54：

原文：防雷装置<u>采用</u>建筑物金属框架做接闪器，引下线<u>采用</u>直径不小于10mm的圆钢。

原译：Lightning arresters adopt metal frameworks of buildings, and down leads adopt round steels no thinner than 10 mm.

改译：Metal frameworks of buildings <u>are used as</u> the receptor of lightning arresters, and down leads are (made of) round steels no thinner than 10 mm.

分析：本例原译都把"采用"译成主动式的 adopt。由于"防雷装置"和"引下线"都不具有主动"采用"的能力，因此原译法是错误的。改译后，将第一个"A采用B"理解成"B被当作A使用"，译成 B is used as A，也可译成 A is B。改译后的第二个"A采用B"，被理解成"A的材料是B"或"A由B制成"，译成 A is made of/from B。

例3.55：

原文：锅炉采用悬吊式结构，全膜式水冷壁炉墙，适当使用柔性膨胀节，以利锅炉的密封性能。

原译：The boiler <u>shall be adopted</u> suspension structure, full membrane water cooling furnace wall, and properly used flexible expansion joint, to facilitate the sealing performance of boiler.

改译：The boiler <u>shall have</u> a suspension structure and a light full-membrane water-cooling wall, with flexible expansion joint if applicable, to improve the sealing performance.

分析：本例的"采用……结构"，是"为/具有……结构"的意思，译文中也不宜出现adopt/use。

例3.56：

原文：布风板及风帽，应选用布风均匀、不易堵塞的型式，并采用优质材料，以保证正常燃烧。

译文：Wind-distribution plate and funnel cap shall have uniform wind-distribution and be hard to block, and made of quality materials, to ensure normal combustion.

分析：本例的"选用"的意思相当于"具有"，"采用"的意思相当于"由……制成"。

例3.57：

原文：本工程升压站主变压器中性点采用经隔离开关直接接地或经避雷器、放电间隙接地。

原译：The method of direct grounding through disconnecting switches or grounding through lightning arresters or discharging gaps is applied for the neutral point of main transformers of the booster station in the Project.

改译：The neutral point of main transformer of step-up station for the Project is earthed with a disconnecting switch directly or with a lightning arrestor or discharging gaps.

分析：本例的"采用"，其意思是"经由""凭借"，表示方式，可以译成with，不必像原译一样译成the method of -ing is applied这样的迂回译法。

例3.58：

原文：进出配电箱/盘的端子，必须加强绝缘并采取固定措施。

原译：Terminals in and out of distribution box/panel must reinforce insulation and take fixation measures.

改译：Terminals in and out of distribution box/panel must be strictly insulated and fixed.

分析：本句的"加强绝缘"是指"被严格绝缘"的意思，"采取固定措施"

是指"加以固定",即"被固定"的意思。因此,原译是明显不妥的。

例3.59:

原文:电缆滑车:用于悬挂电缆的滑动装置,采用整体注塑件制造,按照1.5米距离设置一只。

原译:Cable pulley: a slider used for hanging cables, adopts overall injection-molded parts; it is set every 1.5m.

改译:Cable pulley: a one-piece injection molded slider for hanging cables, arranged every 1.5m.

分析:这里的"采用……制造",不应该把"采用"译出,而是结合整个句子翻译。

小结:

"采用""选用""采取"等词,很多时候不是"采纳"的意思,而是"由……制成""具有"的意思,一般不译成主动式或被动式adopt,而是要译成be made of/ from, have等意思,或者根据语境综合考虑后译出。

(4)结合

例3.60:

原文:厂内道路设计应结合建厂的施工因素,考虑路面分期施工的可能及建厂期间通过超重、超限大件运输的需要。

原译:The design of in-plant roads shall combine the plant construction factors, and consider the possibility of pavement construction by stages and the needs of overweight and oversize transportation during the construction.

改译:The construction factors, including the possibility of pavement construction by stages and the needs of overweight and oversize transportation during the construction of plant, shall be considered in the design of in-plant roads.

分析:本例里的"结合",原译直译成combine,但实际是"考虑"的意思,因此这里将"结合"译成consider。而后面的"考虑",则是具体的考虑因素,因此,可以把后面的"考虑"译成including。

例3.61：

原文：工程施工测量将根据业主提供的坐标点和高程控制点，结合总平面图和施工总平面布置图，建立适合本标段施工的平面控制网和高程控制网。

译文：For engineering construction measurement, a planar control network and an elevation control network appropriate for the construction of this bid section shall be developed based on the coordinate points and elevation control points provided by the Owner and the site plan and general layout.

分析：本例第二小句中的"结合"，与第一小句中的"根据"，意思都相当于"基于"。因此，参考译文将"根据"和"结合"合译成一个based on。

（5）参照、对待、落地/到位

例3.62：

原文：螺旋桨与螺旋桨轴的安装要求，如无专门规定可参照《钢质海船入级与建造规范》（中国船级社，1989年）第三篇轮机第11.5.4条和第11.5.5条的有关规定。

原译：The installation requirements of propeller and propeller shaft, in the absence of special regulations, may refer to the relevant regulations of Article 11.5.4 and 11.5.5 in Part III Turbine in Rules and Regulations for the Construction and Classification of Sea Going Steel Ships (China Classification Society 1989).

改译：The propeller and propeller shaft may be installed, in the absence of special regulations, by following/in accordance with relevant regulations of Article 11.5.4 and 11.5.5 in Part III Turbine in Rules and Regulations for the Construction and Classification of Sea Going Steel Ships (China Classification Society 1989).

分析：本句的"参照……的有关规定"，指的是"遵守……的有关规定"或"按照……的有关规定执行"的意思，不能按字面意思译成refer to。

例3.63：

原文：客服人员应加强与客户的沟通，认真对待客户的投诉与建议，并及时向上级汇报。

原译：The customer service personnel should reinforce the communication with customers, think highly of clients' complaints and suggestions and report to the supervisors.

改译：The customer service personnel should reinforce the communication with customers, seriously deal with their complaints and suggestions and report them to the supervisors.

分析：本例的"认真对待"指的是认真处理投诉与建议，原译法think highly of的意思是"高度评价"。

例3.64：

原文：2016年1—6月，由股权管理团队推进新增落地26笔股票质押式回购业务，初始交易金额合计24.33亿元，延期交易合计5.18亿元。

原译：From January to June in 2016, the quantity of stock pledge buy-back business completed by equity management team is 26. The initial transaction amount adds up to RMB 2.433 billion, and deferred transaction amounts to RMB 518 million.

分析：本句中"落地"是指"落实"，更确切地讲，是指"完成"。

例3.65：

原文：公司注册资本人民币3000万元已经全部到位。

原译：The registered capital of 30 million Yuan has already been in its place.

改译：The registered capital RMB30,000,000 of the Company is wholly paid up.

分析：资金"到位"，指的是已经得到，在这里指的是"已经缴纳"，因此译成be paid up。

（6）国际化/本地化/信息化/智能化/自动化/产业化/商业化

例3.66：

原文：1862年时，英国75%的棉花都来自印度。此时，棉花行业已经走向国际化；埃及和巴西也成了新的供应源。

原译：By 1862, 75% of Britain's cotton originated in India. The industry had become internationalized; Egypt and Brazil also provided new sources of supply.

改译：By 1862, 75% of Britain's cotton originated in India. The industry had gone global; Egypt and Brazil also provided new sources of supply.

分析："国际化"一词，在很多时候指的是产品或行业已经走向全球，因此译成go global，而不是译成become internationalized。类似地，"走向世界"的意思也跟"国际化"一样，也译成go global。

例3.67：

原文：服务外包产业是智力人才密集型现代服务业，具有信息技术承载高、附加值大、资源消耗低、环境污染小、<u>国际化水平高</u>等特点。

原译：Service outsourcing is a knowledge-intensive modern industry, with characteristics of heavy information capacity, high added value, low resource consumption, little environmental pollution and <u>high internationalized level</u>.

改译：Service outsourcing is a knowledge-intensive modern industry, with characteristics of heavy information capacity, high added value, low resource consumption, little environmental pollution and <u>high international collaboration level</u>.

分析：本句的"国际化水平高"这个表达，其实是指服务业外包产业难以由一家公司或一个国家完成，需要国际间的分工合作，也就是国际协作水平很高。

例3.68：

原文：高压电机目前还没有实现<u>本地化</u>，烟台工厂只生产外壳等非铸件产品。

原译：The high-voltage motors had not been <u>localized</u> now, and Yantai factory can only manufacture non-casting products such as the casings.

改译：The high-voltage motors had not been <u>produced locally</u> now, and Yantai factory can only manufacture non-casting products such as the casings.

分析：这里的"本地化"是指在当地生产，而原译的be localized则是指被地方化，被限制在局部或某个地方。

例3.69：

原文：最先进的生产设备的引进和正式投产，预示着研发生产线不断向<u>信息</u>

化和智能化转型。

译文：The introduction and formally putting into production of the most advanced equipment indicated that, the R&D production lines are employing more and more information technology and artificial intelligence.

分析：本句的"向信息化和智能化转型"是指对信息技术和人工智能的利用越来越多，因此，如果字面化翻译成transit toward informationization and intelligentization，会让国外读者看不懂。

例3.70：

原文：渗漏点可通过在污水管中部署自动化机械予以检测。

原译：The leakages can detected by automatic machines installed in the sewage networks.

改译：The leakages can detected by automated machines installed in the sewage networks.

分析：需要注意的是，"自动化（的）"和"自动（的）"这两个形容词意思是不同的，"自动化（的）"一般译成automated，其名词形式为automation，而"自动（的）"一般译成automatic。

例3.71：

原文：产业链完善与创新是国产基础软件产业化的关键。

原译：The improvement of industrial chains and innovation are the key for the industrialization of China-made basic software.

改译：The improvement of industrial chains and innovation are the key for the mass production of China-made basic software.

分析：本句的"产业化"，指的是软件的批量生产，因此译为mass production。而industrialization则是"工业化"的意思，不是"产业化"的合适译文。

例3.72：

原文：藻类生物燃料离商业化还有一定的距离。

原译：There is still some way from the commercialization of algae biofuel.

改译：There is still some way from the commercial production of algae biofuel.

分析：本句的"商业化"，指的是商业化生产，译成commercialization也可以，但显得有点字面化翻译，而译成commercial production则更好理解一些。

二、中国特色名词、形容词及虚词

这里所讲的中国特色名词，多数都是具有中国文化含义的词，而中国特色形容词和虚词，其难译之处，在于国人经常对这些词语的词义理解不到位。请看以下介绍。

（1）我国/国产/国内、中医/国医/国学

例3.73：

原文：我国幅员广大，有着十分丰富的太阳能资源。

原译：Our country has a vast territory, with rich solar energy resources.

改译：China has a vast territory, with rich solar energy resources.

分析：中文讲的"我国"，通常就是指中国，如果译成our country，国外读者可能会误认为是其本国，或者不明白是哪个国家。

例3.74：

原文：国产电视质量已经与日韩等国相差无几。

原译：The quality of domestic TV sets is almost the same as that made in Japan and South Korea.

改译：The quality of China-made TV sets is almost the same as that made in Japan and South Korea.

分析：与上例一样，中文的"国产"，通常是指中国制造，也不能译成domestic或home made，否则会引起理解困难，正确做法是将"国产"译成China-made或后置定语made in China，有时也可译成Chinese或China's。

例3.75：

原文：中医/国医，一般指以中国汉族劳动人民创造的传统医学为主的医学，所以也称汉医。是研究人体生理、病理以及疾病的诊断和防治等的一门学科。

译文：Chinese Traditional Medicine generally refers to the traditional medicine created by the people of ethnic Han, also called Traditional Han Medicine;

it is a science that research the physiology, pathology, and the diagnosis, prevention and control of diseases etc. of human.

分析：根据定义，中医就是中国传统医学，或者汉族传统医学，也称为"国医"。因此，直接译成首字母大写的Chinese Traditional Medicine。

小结：

总之，当碰到"我国""国内""国家""国产"等词时，均应将"我国""国"等词理解成中国，并译成China、China's、Chinese、made in China等。同理，"中医""国医"应理解成中国传统医学，译成Chinese Traditional Medicine，"国学"应理解成中华传统文明研究，译成Sinology（汉学）或studies of Chinese ancient civilization，"国粹"应理解成中华文化精粹，译成the quintessence of Chinese culture。

（2）老/老区/正规

例3.76：

原文：这个曾经是移民库区的老柑橘产业村，随着柑橘品种老化，种植柑橘收益下降，很多人都不愿意种柑橘，外出务工成了主业。

原译：This used to be a village of old citrus industry in relocated reservoir area. With the aging of citrus varieties, the earnings from planting citrus declines, many people are reluctant to produce citruses; hence, working outside becomes the main stream.

改译：This used to be a traditional village of citrus industry in the relocated reservoir area. With the aging of citrus varieties, the earnings from planting citrus declines, many people are reluctant to produce citruses; hence, working outside becomes the main stream.

分析："老柑橘产业村"指的是传统的柑橘产业村，而不是种老柑橘的产业村。所以，"老"要译成traditional。

例3.77：

原文：萩芦镇是莆田市果树老区。

译文：Qiulu Town is an important traditional fruit planting area in Putian City.

分析：本句中的"老区"是指"传统区域"。

例3.78：

原文：陕甘宁<u>革命老区</u>

译文：Shaanxi, Gansu and Ningxia <u>old revolutionary base areas</u>

分析："革命老区"的意思是"老革命根据地"。

例3.79：

原文：本工程的钢筋必须是<u>正规钢铁厂</u>的合格产品，并报建设单位审批后方可使用。

原译：Conforming rebars from <u>regular iron and steel factories</u> must be adopted for the Project, and shall be used only with the approval from the developer.

改译：Conforming rebars from well <u>recognized iron and steel factories</u> must be adopted for the Project shall be used only with the approval from the developer.

分析：这里的"正规钢铁厂"，指的是各种证照齐全，而且运营时间比较长，被社会和业界普遍认可的钢铁厂，所以要译成 well-recognized iron and steel works。

（3）不合格/不符合

例3.80：

原文：项目部定期组织机械设备和手持电动工具安全检查，对<u>不符合</u>填写机械设备和手动电动工具整改单限期整改。

译文：The Project Department should regularly organize safety inspection for machinery and hand-held electric tools, and <u>for the non-conformance item of the machinery and hand-held electric tools, a rectification sheet should be filled out for rectification within a time limit specified</u>.

分析：在程序类文件里，"不符合"可当名词，是"不符合项"的简称，译成nonconformance或nonconformity，是指性能、文件或程序方面的缺陷，因而使某一物项的质量变得不可接受或不能确定。本句的第二分句的意思是"对于不符合项，需填写机械设备和手动电动工具整改单，以进行限期整改"。

而跟"不符合"近义的"不合格",则通常当形容词用,指的是产品不满足要求,应译成non-conforming。而disqualified则是指人"不具有资质""不称职",一般不用于产品。

(4) 较、各、别

例3.81:

原文:对安装难度较大的、较精密的设备要另外编制吊装方案,经公司总工批准后实施。

原译:For relatively accurate equipment that is relatively difficult in installation, the hoisting plan shall be otherwise prepared and put into practice after the approval of the company's Chief Engineer.

改译:The hoisting plan of hard-to-install accurate equipment shall be otherwise prepared and executed with the approval of the company's Chief Engineer.

分析:在应用类中文文件中,经常会碰到"难度较大""较精密"等含有"较"字的表达。这是典型的中国式表达。其实,这里的"较",有时候其意思是比较级,必须译成英文比较级,但有时候其意思并不是比较级,可以忽略不译,比如本例。同理,"较好""较大"等词,很多时候也可以直接译成good和large。

例3.82:

原文:在油气生产过程中,对于各项操作均有明确的作业规程。

原译:During oil and gas production process, each operation shall be provided with specific operation regulations.

改译:During oil and gas production process, all operations shall be provided with specific operation regulations.

分析:根据上下文语境,本句的"各项"操作不是"每一项"操作的意思,而是"所有"操作的意思,因此要译成all operations而不是each operation。同时要注意的是,后面的"规程"也要相应改成复数regulations。

例3.83:

原文:公司品保部负责编写和填写各种质量管理报告和记录。

原译:The QA Department is responsible for the preparation and filling of each quality management report and record.

改译：The QA Department is responsible for the preparation and filling of various quality management reports and records.

分析：本句的"各种"质量管理报告和记录，即可认为指的是"各种各样"，不是强调"每一个"，译成various；也可理解成"所有"而译成all，但译成various似乎更恰当。

例3.84：

原文：项目经理将帮助公司在该国处理好各方面的关系。

原译：The project manager will help the company smoothen the relationships of every party in the country.

改译：The project manager will help the company smoothen the relationships between different parties in the country.

分析：本句的"各方面"的关系，指的是三方或三方以上的两两之间的关系，即不同方的关系，宜译成between different parties，译成of every party属于理解错误，误以为是"属于每一方的关系"。

例3.85：

原文：这两种操作方法各有优缺点。

译文：These two operation methods each have their own strength and weakness.

分析：本例的"各有优缺点"，确实是指两者分别都有优缺点，因此，应译成each，但这里的each是副词，不是形容词。

小结：中文的"各"，在不同语境下，可以有不同意思，在翻译的时候，应当思考一下在上下文的语境下是哪种意思，然后再选用合适的译法。如果无法准确区别"各"字在不同语境下的意思，可到百度百科上查找"各 意思"。

例3.86：

原文：同时，投资地国别安全形势的动态数据对于中国企业的对外投资也至关重要。

原译：Moreover, country-specific realtime security data are also of great importance for overseas investments by Chinese enterprises.

改译：Moreover, realtime security data by country are also of great importance for overseas investments by Chinese enterprises.

分析:"国别"的意思是指针对具体国家的,译成前置定语country-specific或后置定语by country是比较合适的译法。

总结:

以上所分析的中文词,之所以容易出现误解误译,本质上还是对中文词意思理解不够透彻。为便于大家学习,现将以上语境下的出现误解的中文词的意思列表如下:

中文词	中文意思	举例
进行	推动或从事某项工作、动作	进行了技术开发状况的考察=考察技术开发状况 译文:survey the technology development conditions
处理	用在动词之后,做补充说明	对配件做镀锌处理=给配件镀锌 galvanize the fittings
老	传统的	老柑橘产业村=传统的柑橘产业村 traditional village of citrus industry
好	完成	做好后续修补工作=完成后续修补工作 complete the subsequent repair work
较	一定程度	较好=好:good 较高=高:high
最	极、非常	最拿手的本领=极擅长的本领:very skilled capability
各	全部、各个、每个、每一个	对于各项操作均有明确的作业规程=对于所有操作均有明确的作业规程:All operations should have specific operation codes. 填写各种报告=填写各种各样的报告:Fill in all types of reports. 处理好各方面的关系=处理好所有相关方的两两关系:Handle the relationships between all relevant parties. 各有优缺点=分别都有优缺点:All have their own strengths and weaknesses.
别	按……逐个分类/说明	国别人权报告=按国家逐个说明的报告:Human Rights Record by country. 地区别GDP=按地区逐个统计的GDP:GDP by region. 产品别负载=按产品逐个分类的负载:Loads by product.

本节课后练习一

翻译下列句子，注意加下画线部分：

1. 一<u>旦</u>发生事故能及时<u>实施</u>救援。
2. 泄漏故障排除后，<u>需重新进行抽真空</u>。
3. 投资方对各公司<u>进行了技术开发状况的考察</u>。
4. 缺陷产品必须<u>经过</u>修理后才能进入下一工序。
5. 若铝合金的晶粒粗大，可以重新加热<u>予以</u>细化。
6. 在开始施工之前，<u>应先做好预可行性研究工作</u>。
7. （无人机）电池<u>采用</u>卡扣式设计，将电池按如图方向扣入飞行器电池舱中即可。
8. 单层铝板加强筋的<u>固定</u>可采用电栓钉，但应确保铝板外表面不应变形、褪色，固定应牢固。
9. 这种轴承座是<u>选用</u>不锈钢制成的。
10. 这篇论文是<u>结合生产实践</u>写的，具有很高的参考价值。
11. 财产保险金额由投保人参照保险价值自行确定，并在保险合同中载明。
12. 在操作时，宜认真对待机器中出现的任何异常。
13. 国产芯片这几年进步十分明显，尤其是华为公司的麒麟处理器。
14. "中国建造"软实力正在走向世界。
15. 我国陆地面积在世界各国中排名第三。
16. 年度产品别生产量预算表

三、其他中国特色表达

1. "典型性"中国特色表达：

这里所讲的"典型性"中国特色表达，是指较常见的，有固定形式的，但对新译员来说又较难译的一些中国特色表达，比如：

由/为……规定/所规定（的）、除非另有规定/定义、如适用/如有必要/如有/如可能/如允许、以……为准、在用时/不用时、视……而定、所在地/所在园。

以下对上述"典型性"中国特色表达予以举例说明。

（1）由/为……规定/所规定（的）

例3.87：

原文：此外，公司应确保规定用于本系统的相关部件（例如：注射针和针管）及其功能（如电动驱动系统和自动化功能）符合ISO 11608相关部分要求。

译文：Moreover, companies shall ensure that the appropriate components (e.g. needles and containers) and features (e.g. electromechanical drive systems and automated functions) specified for use in the system satisfy the relevant parts of ISO 11608.

分析：A specified by/for B：由/为B所规定的A。

A specified for use in B：规定用于B的A。

例3.88：

原文：切断机不得断切超过该机所规定直径的钢筋，不得断切超过刀片强度和烧红过的钢筋，以保证机械安全。

译文：The shearer should not cut off the bars with the diameter larger than that specified for the shearer or bars harder than the blade or red-heated bars, to ensure the safety of the machine.

分析："该机所规定（的）"这一表达，其意思是指"某规范（为该机）所规定的"，在本句中是指"为该机器（即切断机）所规定的"，所以译成that specified for the shearer。

（2）除非另有规定/定义

例3.89：

原文：除非另有规定，否则本规范中所述材料尺寸均为成品尺寸或完全压实的尺寸。

译文：Unless otherwise specified, all dimensions of materials described in the Specification means the finished or fully compacted dimension.

分析：Unless otherwise specified / unless specified otherwise：除非另有规定。

Unless specified on the contrary：除非另有相反规定。

Unless otherwise agreed：除非另有约定。

例3.90：

原文：除非上下文另有定义，否则本协议所提及的"条"或"附件"均指本协议中的条或附件。

译文：Any reference to an "Article" or an "Annex" in the Agreement is, unless the context otherwise defines, a reference to an Article of or an Annex to this Agreement.

分析：Unless the context otherwise defines：除非上下文另有定义。

（3）如适用/如有必要/如有/如可能/如允许

例3.91：

原文：卖方应向买方提交由船级社签发的确认下列事项（如适用）的船级维持证明。

译文：The Seller shall submit to the Purchaser the Class Maintenance Certificate issued by the Classification Society of the Vessel confirming the following items (if applicable).

分析：如适用：if applicable。

例3.92：

原文：检查送料辊运转有无杂音，如有，应立即查找原因。

译文：Check whether there is noise in the conveying roller, and, if any, find the causes immediately.

分析：本句的"如有"的意思是"有任何（上述事/物）"，通常译成if any，可以当作插入语插入。

例3.93：

原文：为了限制温升梯度，如有必要，在加热的最初几小时可以降低煤油蒸汽的温度。

译文：To limit the temperature rise speed, if necessary, the temperature of kerosene vapor may be decreased in the first hours of heating.

分析：如有必要：if necessary；通常可以当作插入语插入。

如有需要：if needed。

例3.94：

原文：请向税务部门说明贵公司的详细纳税情况，如有可能，请提供本年度财务报表。

译文：Please state the tax payment details of your company to the tax authorities, and if possible, please provide the financial statements of this year.

分析：如有可能：if possible；通常可以当作插入语插入。

例3.95：

原文：浇注垂直构件时，砼也可通过溜槽或（如允许）人孔浇注。

译文：When placing the vertical members, the concrete may also be poured through the chute or (if permitted) the access hatch.

分析：如允许：if permitted；如条件许可：if conditions permit。

（4）以……为准

例3.96：

原文：如两者发生冲突，请以电力行业标准为准。

译文：In case of any conflict between them, the standards of power industry shall govern / prevail.

例3.97：

原文：操作手册内容如与实际产品有所偏差，请以实际产品为准。

原译：The actual product shall prevail if there is any discrepancy between the Operation Manual and the actual product.

改译：For any discrepancy between the Operation Manual with the actual product, please refer to the real product.

分析：将"请以实际产品为准"译成The actual product shall prevail，感觉上不地道。为找出地道译法，可在网络上查"For any discrepancy between the Operation Manual and the actual product"。由搜索结果可判定please refer to the real product的表达比较符合英文行文习惯。

（5）在用时/不用时

例3.98：

原文：当不用此装置时，可将直打杆调至最高或直接卸掉。

原译：When not using the device, the straight beating bar can be adjusted to the highest position or directly dismantled.

改译：When the device is not in use, the straight beating bar can be adjusted to the highest position or directly dismantled.

分析：某物处于"使用（在用）/不用"状态，一般可以译成be in use/ not in use。

（6）视……而定

例3.99：

原文：其中一个重要考量，就是候选人具有或缺乏相关经验，<u>具体视情况而定</u>。

译文：An important consideration is the candidates' experience, or lack of it, as the case may be.

分析："（具体）视情况而定"一般可译成as the case may be，类似的还有as is the case（通常就是这样，实际就是这样）、whatever the case be（不论是何情况）。

例3.100：

原文：在规定设计范围内，即0.5 bar 至16 bar、25 bar 或40 bar（<u>具体视阀门位置而定</u>）之间，阀门应通过小孔口放气的方式对空气的存在产生反应。

原译：Valves shall respond to the presence of air by discharging it through the small orifice at any pressures within a specified design range, i.e., 0.5 bar to 16 bar, 25 bar or 40 bar <u>according to upon the location of the valve</u>.

改译：Valves shall respond to the presence of air by discharging it through the small orifice at any pressures within a specified design range, i.e., 0.5 bar to 16 bar, 25 bar or 40 bar <u>dependent upon the location of the valve</u>.

分析："视……而定"除了译成as ... may be之外，还可译成dependent upon ...

（7）所在地/所在国

例3.101：

原文：政局不稳会影响项目的实施，因此，需对项目<u>所在国/所在地区</u>进行必要的政治风险评估。

原译：Political instability may affect the implementation of the project, therefore, necessary political risks must be assessed for <u>the project's region / country</u>.

改译：Political instability may affect the implementation of the project, therefore, necessary political risks must be assessed for the country/ region where the project locates.

分析："所在地"的意思，是某个事物所处的位置，一般不译成sth's location/region，而是译成the place where sth. Locates。本句的"项目所在国/所在地区"，则宜译成the country/region where the project locates。

2. "非典型性"中国特色表达

这里所讲的"非典型性"中国特色表达，是指不容易归类的一些中式表达。请看以下例子：

例3.102：

原文：九镍钢是一种低碳镍合金钢，其标称组分为9Ni，专用于处理、运输和储存温度低达–196℃（–320℉）的液化石油气的低温或低温压力容器和设备。

译文：The 9% nickel steel is a low-carbon nickel alloy steel with nominal composition of 9Ni, which is particularly used for low-temperature or cryogenic pressure vessels and plants for processing, transportation, and storage of LNG down to –196° C [–320° F].

分析：从网络上对"九镍钢"的介绍可知，"九镍钢"是含镍量为9%的特种钢。另外，在网络上查"9% Ni steel"，得到的结果以9% nickel steel居多，只有少部分是9% Ni steel。因此，可以判定"九镍钢"的英文为9% nickel steel。

例3.103：

原文：设备基础施工：必须在设备到货后，核对土建基础图和设备基础，地脚螺栓尺寸无误，方可进行。土建基础设计中有关预留孔、预埋件如与设备有出入，应按实际情况进行修正。

原译：Construction of equipment foundation: after equipment delivery, check the size of foundation bolts against the civil engineering foundation drawing and equipment foundation drawing before construction. In case of any inconsistency between the civil engineering foundation design on the preformed holes and embedded parts and the system, correction shall be made based on actual conditions.

改译：Construction of equipment foundation: after equipment delivery, check the size of foundation bolts <u>by comparing</u> the civil engineering foundation drawing and equipment foundation drawing before construction. In case of <u>any inconsistency of the preformed holes or embedded parts between the civil engineering foundation design and the actual conditions of the equipment</u>, correction shall be made based on actual conditions.

分析：这里的"核对土建基础图和设备基础"是指把两个基础图进行对比，以确保地脚螺栓尺寸无误。"有出入"是指"有差别"。

例3.104：

原文：助力臂是由伺服主控、伺服电机、传感器、电源模块等组成<u>并由处理器控制，符合人体工程学</u>的物料搬运设备。它具有易操作、精度高、智能化、速度可控及安全可靠等特点。

原译：The booster arm comprises of servo master control, servo motor, sensor, power model and so on, and <u>it is controlled by the processor, which accords with the material handling device of human body engineering</u>. It has the characteristics of easy operation, high precision, intelligence, controllable speed, safe and reliable etc.

改译：The booster arm comprises of a servo master control, a servo motor, a sensor, a power module and so on, <u>it is an ergonomic material handler controlled by a processor</u>. It is characterized by easy operation, high precision, intelligence, controllable speed, safety and reliability etc.

分析：本例原译对"并由处理器控制"的翻译不够顺畅，更重要的是"符合人体工程学的"译得很不好。尤其是对后者的翻译，译员没有弄清楚"符合人体工程学的"有一个专门的英文名可用，而采用自己的译法，明显是中式译文。改译后的译文明显更顺畅更简洁。

例3.105：

原文：当气缸中充入压缩空气时，径向柱塞所产生的推力，通过横梁及滚轮与内曲线导轨相互作用，产生扭矩输出，经行星减速机构减速产生强大扭矩，<u>驱动闸阀的阀杆螺母旋转</u>，<u>使阀杆作升降直线运动</u>，开启或关闭闸阀。

原译：When the cylinder is filled with compressed air, the radial plunger produces thrust. Due to interaction of the beam, roller and inner-curve guide rail, the torque output is produced. After the speed is reduced by the planetary reducer, large torque is produced to drive the stem nut of the gate valve to rotate, so that the stem linearly moves up and down to open or close the gate valve.

改译：When the cylinder is filled with compressed air, the radial plunger produces a thrust, which outputs a torque due to interaction amongst the beam, the roller and the inner-curve guide rail, and generates a strong torque after the reduction of the planetary reducer, which further rotates the stem nuts of the gate valve, and moves the stem linearly up and down to open or close the gate valve.

分析："驱动闸阀的阀杆螺母旋转"里的"驱动……旋转"，实际上就是"使旋转"的意思，可以直接译成rotate，而不是译成drive ... to rotate。同理，"使……作升降直线运动"实际上就是"直线升降移动"，因此直接译成move ... linearly up and down。

例3.106：

原文：物联网技术的应用将城市中的水、电、油、气、交通等公共服务资源以及个人和家庭的各种设施和物品通过互联网有机连接起来，达到全面的物联，形成更透彻的感知和更深入的智能化。

原译：The Internet of Things technology connects the public service resources including water, electricity, oil, gas, transportation, various personal and household facilities and items with the Internet, making the Internet of Things all-round with thorough perception and enhanced intelligence.

改译：The Internet of Things technology connects the public service resources including water, electricity, oil, gas, transportation, various personal and household facilities and items with the Internet, realizing thorough perception and enhanced intelligence.

分析：本句后面的"达到全面的物联"是多余的内容，可以不译，"形成更透彻的感知和更深入的智能化"意思等同于"实现透彻的感知和更深入的智能化"。另外,"将……通过互联网有机连接起来"的意思是"用互联网连接"。

例3.107:

原文:<u>绩效考评的结果实行末位淘汰制</u>,优秀的人有机会获得晋升,不及格人员将由人力资源部组织进行培训,降级、限期改进或者强制淘汰,这种机制对员工既有约束也有激励,使管理人员对做不好工作有急迫感和压迫感,把管理人员置于基层员工的监督之下,从而使其更加有效地工作。

译文:<u>Assessment results will be used in the lowliest place elimination</u> to let excellent employees get promotion opportunity, failed employees to be trained, demoted, improved in limited period or phased out under the organization of HR Department. This mechanism is both constrained and incentive to employees, which enable managers to do the work well with the sense of urgency and sense of oppression, putting the management personnel under supervision of the grass-roots employees, to make them work more effectively.

分析:本句的"绩效考评的结果实行末位淘汰制"不宜字面翻译,其意思相当于"(绩效)考评结果被用于末位淘汰中",而后面"优秀的人有机会……不及格人员将由……"的前面,其实可以加上"使得"一词,因此译文里加了let。

例3.108:

原文:根据该工程的特点,调动<u>强</u>有力的管理人员和技术骨干,组成<u>精干</u>、高效的施工部,全面认真履行合同。

原译:According to the characteristics of the Project, appoint <u>strong</u> management personnel and technical backbones to form a <u>lean</u> and highly-efficient construction department and thus perform the Contract comprehensively and conscientiously.

改译:Appoint <u>excellent</u> management personnel and technical backbones to form a <u>highly capable</u> and efficient construction department based on the characteristics of the Project, to perform the Contract comprehensively and conscientiously.

分析：中文的"强有力"用于指人时，指的是人很优秀，"精干"用于指部门时，指的是"能力强的"，而不是"精益的"。因此，将"强有力的管理人员"译为excellent management personnel，将"精干的"译为highly capable。

四、其他中国特色疑难词

本部分所述的内容，都是具有独特的中国语言文化特点的应用技术领域常用的表达，其中很多都是首次碰到的人不容易译好的。这些表达包括：百年/五十年一遇、取两者中较……者、半墙、250厚/120厚、95砖、公斤力、兼任、三个继续、四分管/六分管、X号线、24口全千兆、停车/开车、四班三倒、车、刀/刀具、一道/二道、三底三面、抱闸、过/欠、使能/禁止、上电/下电、（电力）上网/上网电价、所用电。

请看以下对这些表达的翻译解析。

（1）百年/五十年一遇

例3.109：

原文：The barrage is proposed to be constructed for a length of 1632m approximately to ensure the water way of 1275m (approximately) considering 100 years returning flood of 80,000 cum/sec.

译文：考虑到百年一遇洪水流量80,000m^3/sec，拟建拦河坝长度约1632米，以保证拥有约1275m宽的水道。

分析：100 years returning flood of 80,000 cum/sec：百年一遇洪水流量80,000m^3/sec。

（2）取两者中较……者

例3.110：

原文：球罐与扁球罐——润湿面积等于储罐总表面积的55%或到地上高度为30英尺（9.14米）处的储罐表面积（取两者中较大者）。

译文：Sphere and Spheroids —The wetted area is equal to 55% of the total surface area or the surface area to a height of 30 feet (9.14 meters) above ground, whichever is greater.

分析：whichever is greater：取两者中较大者。
　　　whichever is higher/better/longer：取两者中较高/较佳/较长者。

有时候也把"取两者中较……者"说成"以两者中较……者为准"。

（3）半墙、250厚/120厚、95砖

例3.111：

原文：据推算，250厚C30混凝土的设计抗压能力约为502.5吨/m³。

译文：It is estimated that, the designed compression resistance of 250mm C30 concrete is about 502.5t/cm³.

分析：在机械工程等领域，常用尺寸为毫米（mm）。在工程图纸中，如果不标明单位，其单位都是毫米（mm）。因此，本例的"250厚"就是指250mm厚。

例3.112：

原文：家庭内墙装修可采用半墙，如使用空心砖，则墙体宽度连粉刷在内120厚，其主要优点是自重轻（是一般95砖墙2/3），不会对房屋本身结构带来太大负担，隔音效果更好，因为空心砖里的小孔有隔音功能。

译文：The interior wall of home can be half-brick wall, 120mm-thicked for wall and plastering if hollow bricks are used, and has the advantages of low dead weight (2/3 that of the wall made from common standard bricks), without much burden to the home structure, with good sound insulation for the hollow bricks are sound insulated.

分析：本句的"半墙""250厚"和"95砖墙"，对于很多译员来说都很难理解。这种无法理解的中文概念，一般可以在百度里查找相关解释。这里对以上三个词分别说明如下：

① 在百度里查"半墙 意思"，可以知道半墙（即1/2墙、12墙）是半砖墙，厚度为120mm（即"120厚"，含粉刷厚度）。二砖墙（48墙）、一砖墙（24墙）、四分之三墙（3/4墙、18墙）、四分之一墙（1/4墙、6墙）则分别为480mm、240mm、180mm和60mm。

② 再在百度里查"95砖墙 意思"，可以得知"95砖"就是标准砖，尺寸为240mm×115mm×53mm。因此，"95砖墙"就是用标准砌成的墙。与95砖相对应的是85砖，其尺寸是216mm×105mm×43mm。

③ 最后查"95砖 名称来由"，可以得知"95"是标准砖长度为9.5英寸（即240mm）的意思。而85砖则是指8.5英寸（即216mm）

的砖。

了解以上几个词的意思之后,翻译就不会太难了。

(4) 公斤力

例3.113:

原文:当6公斤压力的聚乙烯管装上10公斤压力的阀门时,管道承压不能超过6公斤。

译文:When PE pipes of 6kg/cm^2 are equipped with valves of 10kg/cm^2, the pressure of pipes must be lower than 6kg/cm^2.

分析:本例的"公斤压力"和"公斤"都是"公斤力"或"千克力"意思,都是一种压力单位,其符号为kg/cm^2,意思是压力每平方厘米多少千克,1kg/cm^2 = 100,000Pa,而1个标准大气压= 101325Pa,因此,通常一公斤力近似等于一大气压。

(5) 兼任

例3.114:

原文:董事长可兼任公司总经理、副总经理或其他高级职务。

译文:The President may concurrently act as General Manager, Vice General Manager or other Officer.

分析:中文"兼任"的意思就是在担任某一职务的同时还担任另一职务,即同时担任。

(6) 三个继续

例3.115:

原文:国家海洋局[①]指出,康菲中国必须继续采取进一步有效措施以执行"三个继续"的要求,即及时地"继续及时确定溢油源、及时隔离封堵溢油源、及时清理已泄漏油污"。

译文:The SOA states that COPC needs to continue taking further effective measures to implement the requirements of the "Three Continues", specifically to "continue to identify oil spill release sources, isolate and seal the oil release sources, and clean up the released oil in a timely fashion."

---

① 机构改革前名称。

分析：本例的背景是，美国康菲公司与中海油合作开发的蓬莱19-3油田于2011年6月发生溢油事故，康菲被指责处理渤海漏油事故不力。因此，中国海洋局于2011年9月15日责令康菲公司加大工作力度，继续彻底排查溢油风险点，彻底封堵溢油源，继续清理海上油污，减轻污染造成的损害。这就是"三个继续"说法的来源。本例原文刚好有在"三个继续"后说明其具体含义，因此可以照着原文的表达译出。如果原文只有带有数字的简略表达，比如"三个代表"，通常必须直接译出其具体意思，最好是在字面翻译后，另外立即译出其具体含义。

（7）四分管/六分管

例3.116：

原文：家居用水管的总管一般用6分管，分管可选用4分管，为了保证出水量，分管也可以使用6分管。

译文：For household water pipelines, generally main pipes are DN20 pipes, and branch pipes DN15; branch pipes may also be DN20 pipes to ensure water yield.

分析：这里的"4分"和"6分"是中国式表达法，通过查百度（查"4分管 6分管 意思"或其他关键词），可以知道分别是4/8英寸和6/8英寸（中文将1/8英寸称为1分）的意思，如果以欧洲表达法，则分别相当于DN15和DN20的公称尺寸，而且，8分 = 1英寸（8/8英寸）。所以，应将"4分"和"6分"分别译成DN15和DN20。如果是1英寸，为方便起见，则称为一寸，同理 $1\frac{1}{2}"$ 称为一寸半。

（8）X号线

例3.117：

原文：24号线（AWG24）线径为0.511mm，芯线为圆形。

译文：The 24# wire (AWG24 wire) is 0.511mm in diameter, with round cores.

分析：按照美国线规，超五类线就是24号线（AWG24），线径为0.511mm，芯线为圆形。对于美国线规，24号线即可以写成24# wire/cable，也可以写成AWG24 wire/cable。英制（BWG: Birmingham Wire Gauge）线规、英制标准线规（SWG: Standard Wire Gauge）、美制线规（AWG: American Wire Gauge）都是线规号越大，导线就越细，而公制线规则以直径大小表示导线尺寸。

第三章 中文理解与表达

（9）24口全千兆

例3.118：

原文：

序号	设备名称	规格型号	厂商	数量	单位	单价	总价
14	交换机	24口全千兆	华为	1	台	3,850.00	3,850.00
15	交换机	24口（主口千兆）	华为	2	台	1,700.00	3,400.00

译文：

No.	Name	Spec.	Mnfr	Qty	Unit	Unit Price	Total price
14	Switch	24 × 1000Mbps	Huawei	1	set	3,850.00	3,850.00
15	Switch	24 ports (1000Mbps main port)	Huawei	2	sets	1,700.00	3,400.00

分析：这里的"24口全千兆"的交换机，不熟悉的译员，在网络上很难找到合适译法。24 × 1000Mbps Switch应译为千兆就是指千兆字节（1000Mbps）。

（10）停车/开车

例3.119：

原文：企业糊树脂装置现阶段处于<u>停车</u>，搬迁选址的阶段，仍具有<u>开车</u>的条件。

译文：The paste resin units of the enterprise is now <u>shut down</u> waiting for site selection for relocation, with the conditions of <u>startup</u> remained.

分析：本句的"停车""开车"分别是指机器的"停机""开机"，因此分别译成shut down和startup。而"紧急停车"则译成emergency stop，其缩写为e-stop。

（11）四班三倒

例3.120：

原文：公司电解工采取<u>四班三倒</u>制工作制：即工作8小时、休息24小时。

译文：The company has four groups of electrolysis workers, they perform duties in three shifts, that is, <u>to work 8 hours and then have a rest of 24 hours in turn</u>.

分析：工厂里的"四班三倒"，实际上就是一个岗位有四个工人，每天只有三个工人上班，每人按照上班 8 小时后休息 24 小时的轮班方法上班，相当于每天都有三个人在上班（即"倒班"），另外一人在休息，因此称为"四班三倒"。这样的倒班方法，可以确保一天 24 小时都有人在岗，确保连续运转。而"三班倒"则是一个岗位三个人，每天都上班，但通常隔一段时间大家一起休息，可以译成 have three groups of works to perform duties in shifts all 24 hours，其他倒班方式也可以按类似方法翻译。

（12）车

例 3.121：

原文：<u>精车</u>好第一个零件后，将零件交付检验站检验，合格后继续生产。

译文：<u>Finish turn</u> the first part finely, then deliver it to the inspection station, and continue the production after it passes the inspection.

分析："精车"是指车工行业的精细车削。如果不了解该概念，可在百度里直接查"精车 意思"，结果如下（2020 年 6 月 1 日检索）：

可以见到"精车"经常和"粗车"一起出现，而且第二条链接说这是车工方面的概念。所以，接着在百度里查"车工 意思"，结果如下（2020 年 6 月 1 日检索）：

点击以上"车工 百度汉语"的链接，得知"车工"是指用车床进行切削的工种及从事这种工作的人。在有道词典里查"车工"，得到 latheman，turner，lathe work 等译文，再查 lathe，其意思为"用车床加工""车削"等；查 turn，有"用旋床旋，把……旋成圆形；车，为……车外圆"的意思。因此，可将"车"译为 lathe 或 turn。

再在有道词典上查"精车"，得到的译文是 finish turn，fine turning 等，因此可以推测其译文可能是 finish turn 或 finely turn 之类的。

（13）刀/刀具

例 3.122：

原文：拆下或更换刀具时，请确保机器已冷却至安全温度，防止灼伤。

译文：When dismantling or replacing the tool, please ensure the machine is cooled to safe temperature to prevent burning.

分析：百度百科里对"刀具"的解释是"刀具是机械制造中用于切削加工的工具，又称切削工具"。根据工厂客户的说明，刀具一般不译成 knife 或 cutter，而是译成 tool。其他相关词汇还有：刀架（tool rest）、刀片/刀身（tool blade）、刀座（tool apron）等。

（14）一道/二道、三底三面、抱闸、过/欠、使能/禁止

例 3.123：

原文：The coating system (EHB6 in AS/NZS 2312-2002) shall consist of a single coat of epoxy zinc primer applied to a minimum dry film thickness of 75 microns followed by two coats of micaceous iron oxide (MIO) epoxy each applied to a minimum dry film thickness of 125 microns.

原译：涂层体系（AS/NZS 2312-2002 中 EHB6）应由<u>一层</u>应用于最小干膜厚度为 75 微米涂层的环氧富锌底漆，和<u>两层</u>应用于最小干膜厚度为 125 微米涂层的云母氧化铁（MIO）环氧树脂组成。

改译：涂层体系（AS/NZS 2312-2002 中 EHB6）应由最小干膜厚度为 75 微米涂层的环氧富锌底漆，以及<u>两道</u>最小干膜厚度为 125 微米的云母氧化铁（MIO）环氧树脂组成。

分析：英文 a coat/ two coats 的专业中文表达是"一道/两道"，而不是"一层/道"。"一道"与"一层"有时相同，有时又不同。一层是指在整个涂布范围内都涂上一次，而一道既可以在整个涂布范围内涂一次，也可以是在局部范围内涂一次。句中的"applied to"的意思是"涂布至（多少厚度）"的意思，这里省去不译。

例 3.124：

原文：钢结构表面：所有外露铁件、钢构件除锈后刷聚氯乙烯萤丹涂料<u>三底三面</u>。

译文：For the surface of steel structures, PVC fluorescent coating shall be adopted to <u>apply three primers and three topcoats</u> on all exposed iron and steel pieces.

分析：本句的意思是要在所有外露铁件、钢构件除锈后，给这些钢结构涂上三道聚氯乙烯萤丹涂料底漆和三道聚氯乙烯萤丹涂料面漆。句中的"三底三面"是指三道底漆、三道面漆，类似的还有"两底三面"等表达。

例 3.125：

原文：<u>抱闸</u>是当电梯<u>轿厢</u>处于静止且马达处于失电状态下防止电梯再移动的机电装置。在某些控制形式中，它会在马达断电时刹住电梯。

译文：A contracting brake is an electromechanical device to prevent the moving of an elevator when the <u>lift car</u> is still while the motor is blacked out. In some control modes, it can stop the elevator in case of outrage of the motor.

分析："抱闸"的英文是 contracting brake，是一种闸门，抱合时为制动状态。电梯"轿厢"的英文则为 lift car，可在网络词典里直接查到。

例 3.126：

原文：家用电冰箱、洗衣机等家用电器没有市电过压保护功能，当电网电压波动较大，电压不稳定时，容易烧坏家用电器，或影响其使用寿命。

译文：Household appliances, such as refrigerators and washing machines, have no overvoltage protection functions, and obvious fluctuation of grid voltage can burn them easily or influence their service lives.

分析：电器的"过压"是指电压过高，英文为 overvoltage，"欠压"（undervoltage）则是指电压不足。类似的说法还有：过流（overcurrent）、欠流（undercurrent）。

例 3.127：

原文：远程使能控制在自动模式下使用，位于车站入口和出口侧。

译文：The remote enable control is operated under automatic mode, and located at the entrance or exit side of the station.

分析：使能也称为启用，是指负责控制信号的输入和输出，英文为 enable，其前缀 en- 的意思为"使"，able 意思为"能够"，合起来就是使能。通俗点说，使能就是一个"允许"信号，只有在发出使能信号时，机器才能转动。与"使能"相对的是"禁止"，其英文为 disable。

（15）上电/下电、（电力）上网/上网电价、所用电

例 3.128：

原文：数控车床上电跳闸故障分析与排除。

译文：Analysis and elimination of power-on tripping fault of numerical control lathe.

分析：电学中的"上电"的意思是打开电源，与之相对的"下电"则是断开电源的意思。上电和下电分别译成 power on 和 power off。

例 3.129：

原文：文章分析了电力供求状况、电价上限和总平均上网电价之间的关系，提出了设定电力市场电价上限的新模型。

译文：The paper analyzes the relationships among supply and demand, price cap and overall average feed-in tariff, and proposes a new model for the price cap of electricity market.

分析：Feed-in tariff 指的是用于加速可再生能源技术发展的政策机制，也指这种政策所定的上网电价。在百度里直接查"上网电价"，即可找到两种译法，但应该判断出 on-grid price 应该是中式英文（把这个组合放到网络上查，结果非常少），feed-in tariff 才是正确译法。

例 3.130：

原文：这两种方案提升了变电站所用电的自动化和智能化水平，为目前变电站所用电改造或者未来的设计和建造提供了可行的借鉴方案。

译文：These two schemes improve the automation and intelligence levels of auxiliary powers of substations, providing a feasible reference for the transformation of auxiliary powers of or the design and construction of substations.

分析："所用电"（也称为"站用电"）是指变电所或变电站自身所需用电，而不是"所使用的电"。

本节课后练习二

请解释以下中国特色疑难词的意思，并说明其译法：

百年/五十年一遇	取两者中较……者	半墙
250厚/120厚	95砖	公斤力
兼任	三个继续……	四分管/六分管
X号线	24口全千兆	停车/开车
四班三倒	车	刀/刀具
一道/二道	三底三面	抱闸
过/欠	使能/禁止	上电/下电

（电力）上网/所用电上网电价

第四章　英文理解与表达

第一节　英文遣词造句注意点

一、英文小词的用法

英文句子理解的难点，不仅在于复杂的语法和句式，还包括某些重要实词，甚至是虚词或小词组的意思或用法。这些虚词或小词组包括 in, of, for, that, which, but for, just, amongst, plus, minus, more than, more... than, rather than, rather ... than, such ... that, such ... as to, as such 等。请看以下分析。

（一）关键性小词

在英文文章里，有些句子看似简单，所有词都认识，句子大概意思也理解，但真要翻译时，却无法把中文逻辑说清楚。这种情况，有很多时候都是我们没理解透个别小词所导致的。以下将针对性列举一些例子，以启发大家注意小词的理解。如：

例 4.1：

原文：The standard applies to high-voltage/low-voltage self-protected liquid-filled and naturally cooled transformers for rated power 50 kVA to 1,000 kVA for indoor or outdoor use having a primary winding (high-voltage) with highest voltage **for** equipment up to 24 kV; a secondary winding (low-voltage) with highest voltage for equipment of 1,1 kV.

原译：本标准适用于室内或室外使用的额定功率为 50 kVA 至 1,000 kVA 的高压/低压自我保护式充液变压器及自然冷却变压器，其具有适用于额定电压为 24kV 设备的最高电压初级绕组（高压）；以及额定电压为 1.1 kV 设备的最高电压次级绕组（低压）。

改译：本标准适用于室内或室外使用的额定视在功率为 50 kVA 至 1,000 kVA 的高压/低压自我保护式充液变压器及自然冷却变压器，其初级绕组（高压）的<u>最高设备电压</u>达 24kV，次级绕组（低压）的最高设备电压为 1.1kV。

分析：本句中的 highest voltage **for** equipment = highest equipment voltage，这是本句的难点。

例 4.2：

原文：Foods are considered to be "**high in**" a vitamin/mineral if they provide a minimum of 30% of the Nutrient Reference Value (NRV) **for** that vitamin/mineral (except sodium) per 100g.

译文：如果每 100 克某食物可提供至少 30% NRV（营养素参考值）的某种维生素/矿物质（除钠外），则该食物就被认为"**富含**"该维生素/矿物质。

分析：本句的 high in 其实是指食物在某方面的含量较高，in 是个介词，high in = rich in。后面的 for that 不是一个词组，for 是"对于"的意思，that 相当于 the 或者 such，是 vitamin/mineral 的定语。

本句相关知识的了解，可以上网搜索，百度百科对 NRV 的说明如下（2020 年 6 月 1 日检索）：

NRV (Nutrient Reference Values)，营养素参考值。营养标签中营养成分标示应当以每 100 克（毫升）和/或每份食品中的含量数值标示，并同时标示所含营养成分占营养素参考值（NRV）的百分比。2008 年 5 月，中国卫生部①颁布的"食品营养标签管理规范"说明：营养素参考值（NRV）是食品营养标签上比较食品营养成分含量多少的参考标准，是消费者选择食品时的一种营养参照尺度。"食品营养标签管理规范"提供了食品标签营养素参考值（NRV），标准营养素参考值是依据我国居民膳食营养素推荐摄入量（RNI）和适宜摄入量（AI）制定的。

例 4.3：

原文：When he measures something, the physicist must take great care to produce minimum possible **disturbance of the system** that is under observation.

---

① 机构改革前名称。

原译：在做测量的时候，物理学家必须很小心地把所观察的系统<u>的干扰</u>降至最低。

改译：在做测量的时候，物理学家必须很小心地把<u>对</u>所观察的系统<u>的干扰</u>降至最低。

分析：本句中 disturbance of the system 的意思不是"系统的干扰"，而是"对系统的干扰"，其中的 of 的意思是"对……"或"对……的"。

例 4.4：

原文：The observation **is made of** some phenomenon.

原译：这一观察是<u>由</u>某一现象<u>组成</u>的。

改译：（我们/人们）<u>对</u>某一现象<u>进行</u>了观察。

分析：本句其实是个倒装句，正常的词序应该是：

The observation of some phenomenon is made.

之所以这样倒装，是因为谓语 is made 很短，放在句尾很不好看。

把以上两例做一比较后，可以发现，前一句的 disturbance of the system 和后一句的 observation of some phenomenon 中的 of，其意思不是"属于……的"，而是"对……"或"关于……的"。

类似的例子还有：

例 4.5：

原文：With accurate timetable observance, electric trains can be operated with shorter turnaround times at terminals, which means that more intensive use can <u>be made of</u> rolling stock than is generally possible with steam power.

译文：由于能准确地按照列车时刻表行车，列车在各车站的周转时间缩短了，因此，电力牵引与蒸汽牵引比较，前者能更充分利用机车车辆。

分析：本例的 be made of 不是一个词组，而是跟 use 在一起，组成 use can be made of rolling stock 的表达，相当于 rolling stock can be made use of。

例 4.6：

原文：Unlike the steam or diesel or diesel-electric locomotive, the electric motive-power unit cannot run anywhere <u>beyond</u> the line or lines equipped with conductors.

译文：不像蒸汽机车、内燃机车或电动内燃机车，电力机车离开导电线路就无法行驶。

分析：beyond 原意是"在……之外"，此处应根据语境灵活译成"离开"。

例 4.7：

原文：Regularly review the suitability of the PQP, and shall undertake a full formal review annually as of the Date of Enterprise and submit the findings to the Engineer within Fourteen (14) days of each review date along with an amended PQP should any amendments be required.

译文：定期审核 PQP 的适用性，在进场之日实施全面正式的审核，将结果连同修改后的 PQP（若要求进行修改）在审核日期后十四（14）天内提交至工程师。

分析：Within XXX days of *** = Within XXX days after ***：***之后XXX天。

例 4.8：

原文：Fit a device outside the chamber capable of measuring with an accuracy of ± 1N up to 100N and with an accuracy of ± 1% thereafter the maximum tensile force applied to each specimen.

原译：在室体外安装一个精度为 ± 1N~100N 且精度为各试样所承受最大拉力 ± 1% 的测力装置。

改译：在室体外安装一个可测量对各样本所施加的最大拉力的测力装置，该装置在 100N 以下的精度为 ± 1N，高于 100N 时的精度为 ± 1%。

分析：本句原译文让人感觉不知所云。这里分析如下：

（1）measure 的对象是 the maximum tensile force，中间隔着两个 with an accuracy of ...

（2）两个 with an accuracy of ... 都是 measuring 的方式状语，补充说明 measuring 的精度。

（3）accuracy of ± 1N up to 100N and with an accuracy of ± 1% thereafter 是两个并列结构，意思是"在 100N 以下精度为 ± 1N，高于 100N 时的精度为 ± 1%"。其中，thereafter = after which = after 100N。

例 4.9：

原文：The National People's Congress authorizes the Hong Kong Special Administrative Region to exercise a high degree of autonomy and enjoy executive, legislative and independent judicial power, including that of final adjudication, in accordance with the provisions of this Law.

译文：全国人民代表大会授权香港特别行政区依照本法的规定实行高度自治，享有行政管理权、立法权、独立的司法权和终审权。

分析：本句是《香港基本法》的英文版和对应中文版里的一个句子。句子的理解难点是 that of final adjudication = power of final adjudication = final adjudication power（终审权），在很正式的英文里，因修辞需要，通常比较忌讳同一词的重复使用和名词过度堆叠，因此本句将 final adjudication power 写成介词短语做定语的结构 that of final adjudication。本句中需要说明的另外一点，就是香港基本法中文版"司法权和终审权"是并列结构，但英文版却是 ... independent judicial power, including that of final adjudication，即把终审权当作司法权的一部分。从法律上讲，终审权确实也是司法权的一部分，《香港基本法》英文用了 including 一词，相当于特别强调香港特区拥有司法权中的终审权。而中文则是直接用"司法权和终审权"的并列表达，虽然字面上无强调，但特别提出，也相当于强调了。

例 4.10：

原文：But Americans have fallen out of love with Chinese stocks. This is partly justified, thanks to a series of scandals such as that of Longtop Financial, a Chinese software firm whose former chief financial officer was found guilty in a New York court of "being reckless in making untrue statements" about its finances.

译文：但美国人已经对中国股票丧失了热情。这在一定程度上是合乎情理的，其原因是一系列丑闻的爆出，比如东南融通公司，这家中国软件公司的前首席财务官被纽约法院判定有罪，罪名是其为该公司"不计后果地编制不实财务报表"。

分析：本例的 that of Longtop Financial = scandal of Longtop Financial，也是介词短语做定语，用法与上一例相同。另外，本例后面的 was found guilty in a New York court of "being reckless in making untrue statements" about its finances，意思相当于 was found guilty of ... in a New York court，实质上也是介词短语做定语，只是这个介词短语比较长。

除了以上例子之外，类似的小词用法还有：

rule of law = law rule：法制

war for profits = profit war：利润战

management of technology = technology management：技术管理（对技术的管理）

degree of weathering = weathering degree：风化程度

degree of freedom = freedom degree：自由度

elongation at break = break elongation：断裂伸长率。

例 4.11：

原文：If both Parties cannot reach an amicable settlement, such disputes shall be finally settled by Arbitration in Singapore, in accordance with the Arbitration Rules of the Singapore International Arbitration Centre ("Rules") for the time being in force, which Rules are incorporated by reference into this Section.

译文：如双方未能达成友好解决方案，则该纠纷可在新加坡通过新加坡国际仲裁中心根据其当时有效的仲裁规则进行的仲裁予以最终解决，该规则已通过引用而并入本条。

分析：这里的 which 感觉很令人费解，如果把 which 看成定语从句的先行词，感觉指代不明，解释不通。实际上，这里的 which 意思跟 such 相当，这一点，在解释较全的词典里可以找到，知道这一点之后，本句就好理解了。

例 4.12：

原文：No machine exists which cannot be made reasonably safe.

译文：没有什么机器不能制造得足够安全可靠。

分析：本句实际上相当于：

No machine that cannot be made reasonably safe exists.

也就是说，which 相当于 that，而且 which 引导的定语从句被置于谓语 exist 之后了。

例 4.13：

原文："Ideally, I'd like to have something in my toolkit with which I could influence the housing market <u>and nothing else</u>," says Henrik Braconier of the FI.

译文：监管会成员亨里克·布拉克里尔说，"理想状态下，我希望自己找到一种工具，只影响房地产市场，而不影响其他方面。"

分析：and nothing else 本意是"而不是别的"，但在本句语境下，应将 influence the housing market and nothing else 译成"只影响房地产市场，而不影响其他方面"。

例 4.14：（as 引导的倒装句）

原文：British businesses are very actively engaged in Hong Kong. The 1000 or so British companies with offices in Hong Kong are evidence of that <u>as are the guests at this dinner tonight</u>, they all recognize Hong Kong's strategic location.

原译：英国公司积极参与香港业务，约1000家英国公司在香港开有办事处，以及今晚赴宴的嘉宾都是明证，他们都认为香港具有战略性地位。

改译：英国公司积极参与香港业务。他们认为香港具有战略性地位，约1000家英国公司在香港开有办事处即是明证，出席今晚宴会的嘉宾也证明了这一点。

分析：as are the guests at this dinner tonight 是 as 引导的倒装句，原文相当于：British businesses are very actively engaged in Hong Kong. The 1000 or so British companies with offices in Hong Kong are evidence of that, as the guests at this dinner tonight are evidence of that, they all recognize Hong Kong's strategic location.

类似的例子还有：

The tax code is in the hands of politicians, as are the planning and rent-control regimes that impede the construction of new homes.

税法以及阻碍新房屋建设的规划权和租金管制权一样，都是由政治家制定的。

例 4.15：

原文：Among the stated purposes of the zone are to implement coastal management goals for protecting and encouraging water-dependent uses of the shoreline, and to assure that limited waterfront areas are reserved for the uses they are uniquely suited for and not pre-empted by uses that can be more appropriately located elsewhere.

原译：设置该区域旨在实行沿海管理目标，保护并发展水域依赖性沿海产业，保证有限的滨水区域能够得到恰当利用，不会被更适合他处的建筑所占据。

改译：该区的规定用途包括实施沿海管理目标，以保护并鼓励海岸线的水域依赖性利用，保证有限的滨水区域被用于其唯一适用的用途，而不被用于更适合于其他地方的其他用途。

分析：among 的用法

> among
>
> *prep.*（后接复数名词或代词或集合名词）
>
> 1. 被……所围绕；在……中间：He found it amongst a pile of old books. 他是在一堆旧书中找到它的。
>
> 2. 其中；包括在内：Among those present were the Prime Minister and her husband. 那些出席者中有首相及其丈夫 .* He was only one amongst many who needed help. 他只是众多需要帮助者之一。
>
> 3. 把（部分）……分给每一成员：distribute the books among the class 把书发给全班。
>
> 4. 在……相互之间：Politicians are always arguing amongst themselves. 政客们总是彼此争论不休。
>
> Among these sayings 可以解释成：这些谚语中

又如：

例 4.16：

原文：Donald J. Trump: Big day planned on NATIONAL SECURITY tomorrow. <u>Among many other things</u>, we will build the wall!

译文：特朗普：明天对（美国）国家安全来说是个大日子。<u>最重要的是</u>，我们要开始建墙啦！

分析：本句的政治背景是，2017 年 1 月 25 日，刚上任五天的美国总统特朗普在推特上宣布要在第二天开建美国与墨西哥之间的边境墙，以防止墨西哥人非法越境进入美国。原文的两个句子都挺简洁的，第一句相当于 Tomorrow is a big day planned on National Security。而第二句 amongst many other things, we will build the wall! 的直译是 "除了很多其他事情外，我们还将建墙"，但参考译文根据政治背景，译成 "最重要的是，我们要开始建墙了"。这是一种非常贴切的意译。

例 4.17：

原文：Compliance with this policy should ensure that employees and representatives of Merlin do not breach any of these laws. As mentioned earlier a more detailed Guidance Note on the Bribery Act can be obtained from the Group Legal Director. In addition, Merlin has developed specific online anti-fraud/anti-bribery training. This training is mandatory for those employees whose roles are most likely to <u>expose</u> them to risk in this area.

译文：对本政策的遵守应确保默林员工和默林代表不违反这些法律中的任何一条。如前所述，详细的《反贿赂法案指导说明》可从集团法务总监处获取。此外，默林已开展具体针对性的在线反欺诈/反贿赂培训。这项培训强制工作角色最可能接触欺诈/贿赂的员工参与。这项培训对于其工作职责在本地区最可能<u>面临本法案制裁</u><u>风险</u>的员工为强制性内容。

分析：expose sb. to 这个词，在词典里查到的意思，基本都是 "使……暴露于"。但 expose 有 "使面临" 的意思，所以，根据上下文，可将 expose them to risk 理解为 "使……面临风险"，并增译成 "面临本法案制裁风险"。

例 4.18：

原文：Levocarnitine may also be known as Vitamin Bt. Products <u>labeled as such</u> may have both D and L racemic forms. Use only Levo-(L-) forms as the D- form may competitively inhibit L- uptake with a resulting deficiency. Studies done in rats and rabbits have demonstrated no teratogenic effects and it is generally believed that levocarnitine is safe to use in pregnancy though documented safety during pregnancy has not been established.

译文：左旋卡尼订也被称为维生素 Bt。<u>贴有左旋卡尼订标签的</u>产品可能同时包含左消旋体（L）和右消旋体（D）。卡尼订只可使用左旋体，因为右旋体会竞争性抵制左旋体的摄入，导致摄入不足。对大鼠和兔子的研究表明左旋体无致畸作用，而且一般认为，左旋卡尼订对孕妇无害，虽然尚未有书面记录表明其对孕妇的安全性。

分析：本例内容具有一定专业性。其中 as such = as indicated by what was just said（前面刚刚说过的），这里可将 products labeled as such 意译成"贴有左旋卡尼订标签的产品"或者"这种产品"。

例 4.19：

原文：Man is a biological organism and <u>as such</u> has needs in common with all animals for food and shelter.

译文：人是生物体，作为生物体的人与动物<u>一</u>样需要食物和住所。

分析：本句的 as such = as such capacity = as biological organism，其本义是 as indicated by what was just said（前面刚刚说过的），但在实际翻译时，需根据上下文情况选择合适的译法。

例 4.20：

原文：If a safety feature was thought to be helpful in reducing injury severity in a particular accident, it was <u>marked as such</u>.

译文：如果安全设备对降低在特定事故中伤害的严重性有所帮助，则在此设备上<u>做此说明</u>。

分析：本句的 as such = as indicated by what was just said。此处应采用意译法。

例 4.21：

原文：Contractor means the person, firm or company identified <u>as such</u> in the Agreement and includes its successors and permitted assignees.

译文：承包商是指协议中被确认为此种身份的人、商号或公司，包括其继任者及其许可受让人。

分析：本句的 as such = as such capacity = as the Contractor。

例 4.22：

原文：SOLU-FLAKE is 99.99% soluble in 10% – 15% HCl acid and <u>as such</u> is an excellent additive for drilling fluids.

译文：SOLU-FLAKE 在 10%–15% 的 HCl 中 99.99% 可溶，因此是钻井液的优良添加剂。

分析：本句的 as such = therefore（因此），但其本义仍是 as indicated by what was just said（前面刚刚说过的）。

类似的例子还有：

例 4.23：

原文：He is a lawyer, and <u>as such</u> formally qualified to express opinions about legal matters.

译文：他是个律师，因此有对法律事务发表意见的正式资格。

例 4.24：

原文：I am not against taxes <u>as such</u>, but disagree when taxation is justified on spurious or dishonest grounds.

译文：我不是反对税收本身，但我确实反对无端的或不当理由的税收。

分析：本句的 as such = in itself of in themselves（就其本身而言）。

类似的例子还有：

例 4.25：

原文：The officer of the law, <u>as such</u>, is entitled to respect.

译文：执法官本身理应得到大家的尊重。

其他例子：

(1) Where custom components and parts are provided, each component/part shall be marked to specifically identify that component/part. Printed circuit card cages are defined as an equipment component and <u>as such</u>, shall be clearly identified as stated within this specification.

(2) The written materials shall be as specified, or if not shall be clearly identified as such on the submittal and shall be the best of their respective kind, free of defects and imperfections and suitable for the service intended and subject to the approval of the Engineer.

(3) Employees maintain their Spanish employment contracts and as such are ruled by Spanish legistration and corresponding Collective Bargaining Agreement.

小结：

以上几例介绍的都是小词的用法，多数都是不太常见的用法，但却是译员必须掌握的基本功。对于这些小词的用法，比较好的掌握方法如下：

第一，用一本权威词典或语法书把介词、连词、代词，包括 that、which 的用法认真复习一遍，充分了解这些词的一些不常见用法；

第二，在碰到相关难句时，查一下词典或网络，看看是不是个别小词有意外用法；

第三，从逻辑或相关背景方面思考，有时候会帮助我们发现某些词或短语的意外用法；

第四，把整个句子放到网络上，有时候可以见到类似用法的句子，得到一定的启发。

（二）英文连词使用注意点

英文中有些连词相互之间不宜搭配使用，请看以下四个例子：

例 4.26：

原文：由于我们的通风是强制进风与排风，所以新风风机与排风风机必须严格同步。

译文：As the ventilation is realized by forced intake and exhaust, the fresh air fans must be strictly synchronous with exhaust fans.

分析：英文表原因的连词（as，because，since 等）和表结果的连词（so，therefore，thus 等）一般不在同一句子里出现。

例 4.27：

原文：虽然信用已成为社会经济迅速发展的催化剂，但是事实表明我国还未建立一个市场信用制度。

译文：Though the credit has become the catalyst for the rapid development of social economy, the facts show that China has not established a market credit system.

分析：英文表让步的连词（though, although）和表转折的连词（but, however 等）一般不在同一句子里出现。

例 4.28：

原文：在面临诸如贫困、饥饿、社会不公以及气候变化等广泛而根深蒂固的社会经济与环境问题时，由于其中有些问题是企业活动造成的，人们越来越要求企业以其独创性为这些问题提出创造性的解决方案。

译文：In the face of broad and deep-seated socio-economic and environmental problems, such as poverty, hunger, social injustice, and climate change, partly a result of corporate activities, companies are increasingly being asked to use their corporate ingenuity for providing innovative solutions to these problems.

分析：英文里表示非穷举的词（如 such as, for example 等），与表示穷举的词（如 etc., and so on）一般不在同一句子里出现。因此译文举例结束后没有加 etc.。

例 4.29：

原文：不合适材料是指任何包含以下成分的材料或由以下成分组成的材料：

产自湿地、沼泽或泥潭的材料；

原木、树桩、树根等植物性物质；

易腐烂、易燃烧物质；

浆液或泥浆；

地表土壤、高有机质黏土或淤泥；

液限大于 50% 的材料；

超过 30% 能够通过 BS 75μm 筛的物质。

译文：Unsuitable material shall be any material which includes or consists of:

material from swamps, marshes and bogs;

logs, stumps, roots and vegetable matter;

perishable or combustible material;

slurry and mud;

surface soil and highly organic clay and silt;

material with a liquid limit greater than 50%;

material with more than 30% passing the BS 75 microns sieve.

综合分析：英文连词的使用要注意以下几点：

(a) as/since/because 等表原因的连词与 so/therefore/thus 等表结果的连词不一起使用；

(b) though/although/even if 等表让步的连词与 but/however 等表示转折的连词不一起使用；

(c) such as/for example 等非穷举的词不与 etc. /et al. 等表示穷举的词一起使用；

(d) 在表示包括、包含、组成的词中，consist of, be composed of, be comprised of, be made up of 等是闭合式的，后面列出包含的所有项目，意思是"由……组成/构成/制成"；mainly consist of ...(主要由……组成)和 mainly comprise ...(主要包括……)是半闭合式的，后面列出主要成分；include, comprise, contain 等是开放式的，后面不必列出所有包含项目。

(三) and 的用法[①]

1. and 连接同时发生的动作，或同时存在的属性、特征等，这时可译为"既/又……又……""一方面……，一方面……""而且"等。

例 4.30：

原文：Air has weight **and** occupies space.

译文：空气有重量<u>而且</u>还会占空间。

例 4.31：

原文："We know that exercise improves health," Dr. Chen says, "<u>and</u> feeling healthier might make people feel happier."

[①] 本部分内容摘自以下链接，但有较大改动：http://edu.sina.com.cn/yyks/2014-08-12/1104430708.shtml：托福语法：and 的七种用法你知道吗（2020 年 6 月 20 日检索）。

译文："我们知道锻炼可以改善健康，"陈博士说，"<u>而</u>感觉更健康又可能会让人感觉更快乐。"

2. and 连接不能同时并存的事物或动作时，不能译成"和"，宜译为"或"。

例 4.32：

原文：The whole machine can be assembled **and** disassembled in two hours.

译文：整台机器可以在两小时内完成组装<u>或</u>拆卸。

例 4.33：

原文：The external pressure coefficients Cpe for buildings **and** parts of buildings depend on the size of the loaded area A.

原译：建筑<u>及</u>部分建筑的外部压力系数 Cpe 取决于荷载面积 A 的大小。

改译：建筑<u>或</u>建筑部件的外部压力系数 Cpe 取决于荷载面积 A 的大小。

3. and 连接两个原形动词，表示后一个动作是前一个动作的目的。

例 4.34：

原文：Let me try **and** do the experiment again.

译文：让我试着<u>再</u>测一次实验。

例 4.35：

原文：He told me that he would come **and** help with our design when his work was over.

译文：他告诉我，他忙完后会过来，<u>以</u>帮助我们做设计。

4. and 前面部分表示原因或条件，后面部分表示结果，这时 and = so 或 so that，译为"因此""所以""从而"等。但是，and 表示的因果关系比 so/ so that 弱，通常表示自然而然的顺承式因果关系。

例 4.36：

原文：People with a syndrome called polydipsia feel excessive thirst **and** drink enormous quantities of water.

译文：得有烦渴综合征的人会感觉极度口渴，<u>因而</u>喝下大量的水。

例 4.37：

原文：Books, newspapers and journals are mainly carried by papers and issued in printed manner, **and** collectively termed printed publications.

译文：图书、报纸和期刊主要以纸张为载体，以印刷方式发行，<u>所以</u>被合称为"纸质出版物"。

有时 and 后面部分表示的是前面部分的原因，这时 and = because，宜译为"因为"。例如：

例 4.38：

原文：With hefty incentives, roughly 40 percent of cars bought in Norway last year were electric, **and** the government plans to phase out diesel- and gasoline-fueled cars in the next decade.

译文：由于挪威政府准备在十年内淘汰柴油与汽油动力汽车，并对购买电动汽车给予丰厚奖励，挪威去年所销售的汽车 40% 为电动汽车。

除了两个分句之间可能有如上的因果关系外，and 连接的两个句子成分之间，前后也可能有这种因果关系。例如：

例 4.39：

原文：Sodium carbonate solution has a great number of hydroxyl radicals **and** it is strongly basic.

译文：碳酸钠溶液中有大量氢氧根离子，**因此**呈强碱性。

5. and 之后的部分，有时具有转折或让步意义，这时 and = though，可译为"虽然""但是""仍然"等。

例 4.40：

原文：Several disadvantages tend to limit the use of hydraulic controls **and** they do offer many distinct advantages.

译文：液压控制**虽**有许多突出的优点，**但**也存在一些缺陷，使其应用范围受到了限制。

例 4.41：

原文：Canny edge detection was first created by John Canny at MIT in 1983, **and** still outperforms many of the newer algorithms that have been developed.

译文：1983年，John Canny 在麻省理工学院提出了 Canny 边缘检测算子，至今，这种检测法的检测效果**仍**比其后提出的许多算法更好。

6. and 有时可以表示一种同位关系，这时应将 and 译作"即""也是"。

例 4.42：

原文：When correcting, pay close attention to the seventh **and** last paragraph in the translation.

译文：改稿时，请密切注意译文的第七段，**即**最后一段。

例 4.43：

原文：This is the last **and** most complicated link of the manufacturing process.

译文：这是制造工艺的最后一个环节，**也是**最复杂的环节。

7. and 用在数字或数学概念中，表示相加或相乘。

例 4.44：

原文：Six **and** seven is thirteen.

译文：六**加**七等于十三。

例 4.45：

原文：If the product of A and B is one, then A and B are the reciprocal of each other.

译文：如果 A **乘** B 的积等于一，则 A 和 B 互为倒数。

分析：在带分数中，and 也表示相加。翻译时要按汉语中数字的习惯说法处理，并注意 half, quarter 等的译法。例如：

例 4.46：

原文：On the average, oceans are two **and** one third miles deep.

译文：海洋的平均深度为二**又**三分之一英里。

分析：在 "between ＋数字＋ and ＋数字" 的短语中，and 等于介词 to，译为 "到" 或 "至"。例如：

例 4.47：

原文：The lamp lights when the transistor temperature probe is connected to the rear panel and its temperature is between -20℃ **and** 60℃.

译文：当晶体管测温器接到后板上并且温度在 -20℃**到** 60℃之间时，这盏灯就亮了。

例 4.48：

原文：Tests were performed in stirred tanks, with volumes between 5 **and** 50 litres.

译文：试验是在一些搅拌槽中进行的，槽的容积介于 5 升**至** 50 升之间。

8. and 用在强调结构中。

（1）and 可连接同一个形容词的比较级，表示发展和程度的变化，例如：

例 4.49：

原文：The patient is getting worse **and** worse.

译文：患者的病情在<u>不断</u>恶化。

例 4.50：

原文：The temperature became lower **and** lower with the increase of height.

译文：随着高度的增加，温度越来越低。

（2）and 可连接同义词，表示同一意思，近义并列取其同义。例如：

例 4.51：

原文：Hydrogen can **thoroughly and totally** react with oxygen.

译文：氢气可以与氧气发生<u>完全</u>反应。

例 4.52：

原文：The project must **meet and satisfy** all the terms and conditions of the employer.

译文：该工程必须<u>满足</u>业主的所有条款条件。

例 4.53：

原文：The plan has been identified as a **solid and stable** operation strategy.

译文：这一计划已经被确认为<u>稳妥的</u>经营策略。

（3）重复陈述已经表达过的内容，表示特别强调。例如：

例 4.54：

原文：**We need to know and we need to know** as much as possible in order to adapt to the sort of changes in all aspects of science that will fall upon us like a tidal wave.

译文：为了适应各个科学领域发生的巨大变化，<u>我们非常需要掌握</u>大量的知识。

二、独立主格结构[①]

独立主格结构有很多种形式，可在句子中充当时间、条件、原因状语，或者表示伴随状态或补充说明。独立主格结构主要包括与主句联系紧密的由 with/

[①] 本部分内容有参考百度百科"独立主格结构"词条，其链接为：https://baike.baidu.com/item/%E7%8B%AC%E7%AB%8B%E4%B8%BB%E6%A0%BC%E7%BB%93%E6%9E%84/87655?fr=aladdin（2020 年 6 月 20 日检索）。

without 引导的独立主格结构，以及与主句关系松散的一般独立主格结构。

例 4.55：

原文：If postweld heat treatment is specified by the Welding Procedure Specification, all PQR testing shall be done with the test weldment in the postweld heat treatment condition.

译文：如果《焊接工艺规范》规定需进行焊后热处理，试验焊件的所有工艺评定记录（PQR）试验应在试验焊件处于焊后热处理状态下进行。

分析：本例后面部分有一个 with 引导的独立主格结构，表示伴随状态。with/without 引导的结构在中译英的时候可以经常使用。其中 PQR = procedure qualification record（工艺评定记录）。

例 4.56：

原文：Almost all metals are good conductors, silver being the best of all.

译文：几乎所有的金属都是良导体，其中银的导电性最好。

分析：本例的后半句是"名词+-ing"形式的一般独立主格结构,表示伴随状态。

例 4.57：

原文：These awards will be given to individuals who have made great contribution in related field, ideally each being measurable by numeric criteria.

译文：这些奖励将授予在相关领域做出突出贡献的个人，最好是每个贡献都可以通过量化的标准进行衡量。

分析：本例的后半句是 each 引导的强调型独立主格结构，是对前面 contributions 做补充说明。

小结：

1.各种独立主格结构的形式区别：

（1）with 引导的独立主格与主句逻辑关系紧密，其形式有以下几种：

with + n. + -ed/-ing 形式

with + n. +adj.

with + n. + 介词短语

with（without）+ 宾语（名词/代词）+ 宾语补足语，宾语通常由名词或代词充当，但代词一定要用宾格。

（2）一般独立主格结构与主句的逻辑关系较为松散，其主要形式包括：

名词/主格代词+分词（-ing、-ed）形式

名词/主格代词+不定式

名词/主格代词+介词短语

名词/主格代词+形容词

名词/主格代词+副词

名词/主格代词+名词

（3）each 引导的独立主格结构形式：

句子+复数名词结尾, each+介词短语/形容词短语/名词短语/-ing 形式/-ed形式

（4）There being/It being + 名词（代词）

2. 独立主格结构的特点：

（1）独立主格结构的逻辑主语与句子的主语不同，它独立存在。

（2）名词或代词与后面的分词、形容词、副词、不定式、介词等是主谓关系或动宾关系。

（3）独立主格结构一般有逗号与主句分开。

本节课后练习

翻译下列句子，注意英语句子的下画线部分与汉语句子的连词：

1. This method can reduce the disturbance <u>of</u> the system under observation.
2. Chengdu also wants to promote a favourable view <u>of</u> China, he notes, and that is far easier to achieve with startups than with established multinational companies.
3. Unlike the steam or diesel or diesel-electric locomotive, the electric motive-power unit cannot run anywhere <u>beyond</u> the line or lines equipped with conductors.
4. A relatively large percentage of nonmetallic impurities serves, <u>among</u> other features, to distinguish wrought iron from steel.
5. The German carmakers BMW and Daimler said on Wednesday that they had taken action against executives <u>involved in</u> an organization that sponsored emissions experiments on monkeys, as the companies tried to squelch a public outcry that threatens to tarnish the image of Germany's most important exports.

6. The shares <u>ended up</u> by about half a percent in trading on the Tokyo Stock Exchange on Thursday.
7. The major is serving without pay, <u>as are</u> the city's two other elected officials.
8. The northeastern U.S. and eastern Canada have also been affected by this form of air pollution, and other areas of the two countries <u>are also showing</u> increasing signs of damages, <u>as are</u> other regions of the world.
9. If the computer cannot be shut down, turn off the power, <u>which</u> method is the final means.
10. No machine exists <u>which</u> cannot be made reasonably safe.
11. A rotating wheel will be stopped by a torque such as <u>that</u> of friction.
12. The process of oxidation in human body gives off heat slowly **and** regularly.
13. If a body is acted upon by a number of forces **and** still remains motionless, the body is said to be in equilibrium.
14. After firing the torpedoes the forward part would lose weight **and** the submarine would be out of balance.
15. Reproduction provides new generations **and** makes possible the continuation of race.
16. Aluminum is used as the engineering material for planes and spaceships **and** it is both light and tough.
17. Put the electromagnet in place **with** the end about 1/4 inch above the iron bar.
18. The orbit of each planet is an ellipse, <u>with</u> the sun being at one focus.
19. The water flows away, <u>without</u> leaving any trace.
20. The manometer, <u>being</u> newly repaired, ought to be calibrated.
21. The trouble removed, the motor work smoothly again.
22. After the restructuring, the group will be divided into three subsidiaries, <u>each</u> with its own distinct features.
23. <u>Unless</u> specified otherwise at the time the order is placed, the compressor must run normally under the environmental conditions indicated below.
24. Even with modern anaesthetic techniques not all male patients are fit for this surgery.

25. If the occurrence of force majeure severely hinder the performance of the Contract, either party may terminate the Contract in advance.
26. The fittings, valves and connections of all pipes satisfy the requirements of working pressure of the pipelines.
27. 因为电阻率是温度的函数，所以同一电阻丝在不同温度下有不同的电阻值。
28. 虽然风电项目开发规模已经得到压缩，但首期开发规模依然达到 200 兆瓦。
29. 在市场经济发育不成熟的发展中国家，经常会出现诸如假冒伪劣、有毒有害食品非法生产经营等食品安全问题。
30. 不管包含该函数的行是否被执行，开始时间、结束时间和间隔计时器均应及时更新。

第二节　词义辨析

一、一词多义

apply, component, function, normal, plan, power, product, reduce, terminal, where

apply

例 4.58：

原文：These technical specifications shall apply to all the works as are required to be executed under the contract.

译文：这些技术规范应适用于本合同项下要求执行的所有工程。

分析：apply to 的意思是"适用于"，指的是某种规章制度可被应用于某种情形。

例 4.59：

原文：In materials science, the strength of a material is its ability to withstand an applied load without failure or plastic deformation.

译文：在材料学中，材料强度是指该材料在不产生失效或塑性形变的情况下抵抗所施加的荷载的能力。

分析：在与力相关的上下文中，apply 通常是"施加"的意思，这时经常与 apply 搭配的词为：force, load, stress, pressure 等。

例 4.60：

原文：Thicker coats can be applied if more elastomeric properties are desired.

译文：如想得到更好的弹性，可将涂层涂布得更厚一些。

分析：在油漆涂料相关行业中，apply 的意思是"涂布"，这时经常与 apply 搭配的词为：coating，paint，adhesive，ink 等。

component

例 4.61：

原文：All components shall be carefully inspected before assembly.

译文：在组装前，所有部件均应进行详细的检验。

分析：在机械设备相关的上下文中，component 通常是"部件""组件"的意思。

例 4.62：

原文：Iron is a critical component of hemoglobin.

译文：铁是血红蛋白的关键组分。

分析：在生物、化学、医学、药学等领域中，component 通常是"组分""成分"的意思。

例 4.63：

原文：The force acted on a car can be decomposed into a vertical and a horizontal component.

译文：施加在汽车上的力可以被分解为垂直方向和水平方向的两个分量。

分析：在与力学相关的上下文中，component 通常是"（力的）分量"的意思。

function

例 4.64：

原文：Four major functions of management: planning, organizing, leading, and controlling.

译文：管理的四大功能：规划、组织、领导与管控。

分析：本句中 functions 的意思是"功能"。

例 4.65：

原文：The demand for housing is more properly considered to be a function of income averaged over a period of years rather than just current come.

原译：住房需求更应该被看作是数年期间的平均收入而不是仅仅是当前收入

的职能。

改译：住房需求更应该被看作是数年平均收入的函数，而不是仅仅看作是当前收入的函数。

分析：本例的 function 对很多新译员来说都是很难理解的一个词。但如果认真地到词典里查 function 的解释，就可以发现"函数"这个词义还是比较容易查到的。

normal

例 4.66：

原文：The normal temperature of human body is about 37℃.

译文：人体正常温度约为 37℃。

分析：本句中 normal 的意思是"正常（的）"。

例 4.67：

原文：A line normal to a curve at a given point is the line perpendicular to the line that's tangent at that same point.

译文：曲线在给定点的法线是指该点切线的垂直线。

分析：本句中的 normal 意思是"法向的""垂直的"，normal line 意思是"法线"。在几何学中，法线是指在某一点与该点所在直线或所在平面相垂直。

例 4.68：

原文：Data can be "distributed" (spread out) in different ways. But there are many cases where the data tends to be around a central value with no bias left or right, and it gets close to a "Normal Distribution".

译文：数据的"分布"形式多种多样。但在很多情况下，数据倾向于在中心值左右毫无偏斜地分布，使其分布接近于"正态分布"。

分析：normal distribution 是统计学中的"正态分布"，即"正常状态的分布"。

plan

例 4.69：

原文：A police officer prepared a perfect plan.

译文：有位警官制订了一个完美的计划。

例 4.70:

原文: The Seller has applied for and obtained preliminary and final site plan.

译文: 卖方已经申请并获得初步平面图及最终总平面图。

power

例 4.71:

原文: The power of this motor is 10kW.

译文: 这台电机的功率是 10kW。

例 4.72:

原文: The power is this press is off.

译文: 这台压机电源已经关了。

例 4.73:

原文: The sixth power of 2 is 64.

译文: 2 的六次幂是 64。

例 4.74:

原文: This amplifier has a power of 10.

译文: 这个放大器的放大倍数为 10。

product

例 4.75:

原文: The main problem of this product is that its service life of battery is short.

译文: 这个产品的主要问题是其电池寿命太短。

例 4.76:

原文: Surface pressure acts on the seal area of the tubing, producing an upward force equal to the product of the pressure and the cross-sectional area of the tube.

译文: 地面压力作用于油管密封区，所产生的向上的力与压强和油管横截面的乘积相等。

reduce

例 4.77:

原文: Electric vehicles will reduce the dependency on oil, improve the global energy efficiency and reduce the total CO_2 emissions from road transportation if the electricity is produced from renewable sources.

译文：使用可再生能源电力的电动车辆能够降低对石油的依赖，提高全球能源利用率，并减少道路交通中的二氧化碳排放总量。

分析：本句中 reduce 的意思是其基本词义"降低""减少"。

例 4.78：

原文：The equation can be reduced into a simpler one.

译文：该方程可以简化为更简单的方程。

分析：本句中 reduce 的意思是"简化"。

例 4.79：

原文：The slag may also be reduced in a separated furnace.

译文：炉渣也可以在独立的炉子里还原。

分析：在冶金、化学、生物等领域中，当涉及化学反应时，reduce 是"还原"的意思。

terminal

例 4.80：

原文：Hilton Beijing Capital Airport is close to No.3 terminal of Beijing Capital Airport, with many golf clubs conforming to international standards located nearby.

译文：北京首都机场希尔顿酒店毗邻北京首都机场三号航站楼，周边更汇集了多家国际标准的高尔夫俱乐部。

分析：本处的 terminal 是指机场的航站楼。

例 4.81：

原文：Grand Central Terminal is a commuter, rapid transit, and intercity railroad terminal at 42nd Street and Park Avenue in Midtown Manhattan in New York City, United States.

译文：纽约中央火车站是通勤、快速运输及城际铁路火车站，位于美国纽约市曼哈顿中城第 42 街和公园大道的路口。

分析：本句的两个 terminal 都是火车站或公交车站的意思。

例 4.82：

原文：Ishigaki Port Ritoh Terminal.

译文：石垣市石垣港离岛码头。

分析：terminal 也可指港口的码头。

例 4.83：

原文：A "click" should be heard when the terminal is pluged into the connector.

译文：当将端子插入连接器的时候，应可听到"咔嚓"声。

分析：terminal 在电器中指接线端子。在较大的设备中，terminal 还可当"终端设备"解释。

where

例 4.84：

原文：The maximum fender spacing shall be calculated by the following formula:

$$S \leq 2\sqrt{r^2-(r-h)}$$

Where:

S = Max. fender spacing;

R = Bow radius;

h = Fender height in rated compression deflection

译文：护舷最大间距一般按下式计算：

$$S \leq 2\sqrt{r^2-(r-h)}$$

式中：

S = 护舷最大间距；

r = 船首半径；

h = 护舷被压缩到设计压缩量时的高度。

分析：在公式、表格、图片等的说明中，where 的意思是"其中""式中""表中""图中"。

例 4.85：

原文：Where the Board of Directors determines that the dividend shall be paid by the distribution of specific assets, the Board of Directors may settle all problems concerning such distribution.

译文：如果董事会认为股息应以特定资产支付，则董事会应解决好与股息分配相关的所有问题。

分析：在法律英文中，可将 where 当作 if 使用，引出条件从句，这种用法非常普遍。

小结：

我们在学校教材里学到的词的意思，通常都只是最基本词义。只有在翻译实践中，才能充分体会到英文词义的丰富性。以上举例的只是一小部分多义词，下表列出了更多的多义词供大家学习理解：

英文单词	中文意思及举例
account	*n.* 账户：bank account（银行账户） *n.* 客户：account manager（客户经理） *n.* 原因：on this account（因此；于是）
common	*adj.* 普通的/常见的：common trees（常见树木） *adj.* 公共的：common space（公共场所） *adj.* 共同（的）：common interests（共同兴趣） *adj.* 共用（的）：common switch（公共开关）
complete	*adj.* 完整/整个（的）：a complete machine（整个机器） *adj.* 齐备的，具有：a house complete with furniture（家具齐备的房子） *vt.* 填写：complete the table（填写表格）
device	*n.* （电子）设备：peripheral devices（［电脑等的］周边设备） *n.* （医学）器械：medical device（医疗器械） *n.* （机械）装置、器：locking device（锁定装置）；sensing device（传感器）
edge	*n.* 边缘；缘：edge of extinction（灭绝边缘） *n.* 刀锋；刀刃：edge of sword（剑锋；剑刃） *n.* 优势：an edge over the opponent（与对手相比的优势） *vt.* 缓慢移动：edge toward the exit（朝出口挪动）
form	*n.* 形式、格式：in the form of（以……形式/格式） *n.* 表、表格：application form（申请表、申请书）
matrix	*n.* 矩阵：dot-matrix（点矩阵）；conjugate matrix（共轭矩阵） *n.* 基质：bone matrix（骨基质）；cell matrix（细胞基质） *n.* （地质）基岩：matrix stress（基岩应力）；matrix porosity（基岩孔隙度）

（续表）

英文单词	中文意思及举例
operation	*n.* 操作：manual operation（人工操作）；computer operation（计算机操作） *n.* 作业：drilling operation（钻井作业）；logistics operation（物流作业） *n.* 经营、运营：sustainable operation（永续经营）；commercial operation（商业运营） *n.* 运算：algebraic operation（代数运算）；arithmetic operation（算术运算） *n.* 作战：autonomous operation（独立作战）
organ	*n.* （动物或人体）器官：human organs（人体器官） *n.* （组织）机关、机构：state organ（国家机关） *n.* （音乐）风琴：pipe organ（管风琴）
practice	*n.* 实践：teaching practice（教学实践） *n.* 惯例：best international practice（最佳国际惯例）
prepare	*vt.* 准备、预备：prepare food（准备食物）；prepare a room（预备房间） *vt.* 制备：prepare a solution（制备溶液）；preparing technology（制备工艺） *vt.* 制订：prepare a plan（制定计划）；prepare a law（制定法律）
scale	*n.* 规模：investment scale（投资规模） *n.* 比例尺：map scale（地图比例尺） *n.* 水垢：scale buildup（结垢） *n.* 称/磅秤：platform scale（台秤） *n.* 刻度：scale plate（刻度板、分度板） *n.* 鳞；鳞甲：scales of fish（鱼鳞）
sign	*vt.* 签署（文件）：sign a contract（签署合同） *n.* 标志，记号：traffic sign（交通标志） *n.* （数学）符号：mathematical sign（数学符号）；signed value（带符号值）
signal	*n.* 信号；标志：stress signal（遇难求救信号） *vt.* 标志（着）；表示：the submarine signaled for help（潜艇发出求救信号） *vi.* 发信号
step	*n.* 阶梯、踏板：step of a car（汽车踏板） *n.* 级、档、步进：stepless reducer（无级减速器） *vt.* 踩、踏步：stepping motor（步进电机） *n.* 阶跃：step function（阶跃函数）

（续表）

英文单词	中文意思及举例
switch	*n.* 开关：pull switch（拉线开关） *n.* （通信）交换机：access switch（接入交换机） *n.* （铁路）道岔：runaway switch（安全道岔） 注：switch 本意是"转换"，开关、交换机和道岔本质上都是一种转换工具
term	*n.* 期限；任期：term of service（任职期；服役/使用期）；term of office（任期）；term of validity（有效期） *n.* 条款：terms of delivery（交货条款）；terms of payment（付款条件） *n.* 术语：medical term（医学术语） *vt.* 将……称为：an accident termed as a tragedy（被称为悲剧的事故）

二、近义辨析

第一组词：

staff, personnel, employee, person

例 4.86：

原文：医院全体医务人员都必须通过相应的资格考试。

译文：The medical staff must past corresponding qualification examinations.

例 4.87：

原文：出现异常情况，如设备中有异物，设备异响，局部变形，安全装置破损等，应立即停止运转，并让相关人员撤离操作场所及周围环境。

译文：In case of any abnormity, for example, foreign bodies in, abnormal noise or local deformation of equipment, or damage to safety devices, stop the operation immediately, evacuate relevant personnel from the operation site and the surroundings.

例 4.88：

原文：截至 2015 年底，华为全球员工总数约为 17 万。

译文：The total number of employees of Huawei in the world was about 170,000 at the end of 2015.

例 4.89：

原文：本文件不构成协议，也非业主向潜在申请人或任何其他<u>人员</u>提供的报价或邀请函。

译文：This document is not an agreement and is neither an offer nor invitation by the Employer to the prospective Applicants or any other <u>person</u>.

综合分析：以上几例出现了"人""人员""员工"等词，但参考译文译法各不相同。这里区别如下：

单词	意思	说明
staff	（全体）员工/职员	复数名词，表示整个组织机构或整个部门的全体员工，不得出现 one staff、two staffs 的表达。
personnel	人员	复数名词，表示多个人员。
employee	员工，雇员	可数名词，表示普通的员工。
person	人，人员	可数名词，表示普通的人或人员。

除此之外，crew, people, audience, committee 的用法也跟 staff, personnel 类似，即不能理解成单个或可数的几个人，无复数形式。

第二组词：

dedicated，occupational，professional，special，specialized

例 4.90：

原文：本项目配备有一些<u>专用电脑</u>。

译文：Some <u>dedicated computers</u> are provided for this project.

例 4.91：

原文：<u>职业</u>健康安全评估体系（OHSAS）

译文：<u>Occupational</u> Health and Safety Assessment System (OHSAS)

例 4.92：

原文：专业技术是高校毕业生在专业实践中获得成功所必须掌握的技能。

译文：Professional skills are skills that graduates need to posses in order to be successful in their professional practice.

例 4.93：

原文：仪表管的截断和弯曲应使用专用工具进行。

译文：The instrument pipes must be cut and bent by special tool.

例 4.94：

原文：焊工在管道内焊接时，必须有专人监护，监护人必须熟知焊接操作规程和抢救方法。

译文：A special person must be appointed to guard the welders operating inside the pipeline, and must be familiar with the welding operation procedures and first-aid methods.

例 4.95：

原文：必须组织现场操作人员学习熏蒸消毒专业知识、规章制度……

译文：Onsite operators must be organized to learn the specialized knowledge and rules and regulations for fumigation, ...

综合分析：以上几例都是与"专用""专业""专门""职业"等相关的词。下表是对这些词的区别的补充说明。

单词	意思	说明
dedicated	专用的，专设的	表示专门用途，专为某目的而设。如：passenger dedicated line（客运专线）
special	专门的，专用的，特别的	表示特别为某目的而设。如：special service（专门服务）
specialized	专业的	表示具有某种技术性，具有针对性的。如：specialized LCD producer（专业的液晶显示器制造商）
occupational	职业的，与职业相关的	表示与某职业相关。如：occupational disease（职业病）
professional	专业的，职业的	表示与某个专业相关。如：professional golfer（职业高尔夫球员）

第三组词：

make，model，type

例 4.96：

原文：Vauxhall (make) Corsa Activ (model) 5-door hatchback (type).

译文：沃克斯豪尔（品牌）科莎（型号）五门掀背车（型式）。

分析：在中文里，make和brand这两个词的译法不好区别。make这个词几乎只用在汽车行业，是"品牌/牌号"的意思，Vauxhall这个make是通用汽车公司在英国所售欧宝汽车的一个品牌。model是型号，指的是车的款式，型号的名称可以称为brand（牌子/牌号）。而type则是型式，指的是车的具体形状、大小等。

此外，钢材的牌号和化工产品的牌号，其意思相当于"等级"，所对应的英文是 grade。

第四组词

test/testing，inspect/inspection，check/checking，detect/detection，examine/examination，experiment

例 4.97：

原文：Test methods for measurement of tyre sound emission—drum method

原译：轮胎室内噪声实验方法——转鼓法

改译：轮胎室内噪声试验方法——转鼓法

分析：本例原文是中国一个国家标准的英译官方名称，改译译文是该标准的中文官方名称。原译把"test"译成"实验"，但实际上，英文 test 在多数情况下都与中文"试验"相互对应。

例 4.98：

原文：Both South Korea and Japan are performing out-of-pile oxidation experiments with several graphite materials and SiC TRISO coated fuel particles under the air ingress accident condition for high-temperature gas-cooled reactor.

译文：韩国和日本都正在为高温气冷反应堆进气事故条件下，用数种石墨材料和碳化硅三结构同向(SiC TRISO)包覆燃料颗粒进行堆外氧化实验。

分析：experiment 一般译成"实验"。英文 experiment 基本与中文"实验"相互对应。

《现代汉语词典》对这"试验"和"实验"这两个词的定义如下：

【实验】为了检验某种科学理论或假设而进行某种操作或从事某种活动。

【试验】为了察看某事的结果或某物的性能而从事某种活动。

虽然 test 当名词时，多数情况下译成"试验"，但在无损检验（也称为"无损检测"，英文 non-destructive test）领域中，各种 test 一般译成"检验"或"检测"。如：

涡流检验/检测（ECT）—eddy current test

射线照相检验/检测（RT）—radiographic test

超声检验/检测（UT）—ultrasonic test

磁粉检验/检测（MT）—magnetic particle test

但是，这些无损检验英文里的 test，也可用 inspection 或 examination 替代，意思相同。

以上两例是关于 experiment 和 test 的名词用法的区别。以下介绍 test 的动词用法与其他相关动词的区别：

例 4.99：

原文：The Manufacturer shall sample and test the concrete on a routine basis as required BS 5911, unless otherwise approved by the Engineer.

原译：除非经工程师另行批准，否则制造商应根据 BS 5911 的要求对混凝土进行常规抽样并实验。

改译：除非经工程师另行批准，否则制造商应根据 BS 5911 的要求对混凝土进行常规抽样并试验。

分析：test 在自然科学领域中，最常见译法是"试验"和"测试"，其中"试验"的译法更多。

例 4.100：

原文：It is the responsibility of the receiving agency to inspect all materials, supplies, and equipment upon delivery to ensure compliance with the quality requirements and specifications.

原译：收货代理人应在交货后负责检查所有材料、供应品和设备，以确保其符合质量要求和规范。
改译：收货代理人应在交货后负责检验所有材料、供应品和设备，以确保其符合质量要求和规范。
分析：inspect/inspection 一般都译成"检验"，较少有其他译法。"检验"是指以某种标准对受检对象进行验证，包括检查和验证两个方面。

例 4.101：

原文：Check the oil valves for leakage and seepage and repair or replace the defective valves if necessary to eliminate any fire or explosion risk.

原译：检验油阀是否泄漏，修理有问题的阀门，必要时予以更换，以消灭任何火灾或爆炸隐患。

改译：检查油阀是否泄漏，修理有问题的阀门，必要时予以更换，以消灭任何火灾或爆炸隐患。

分析：本句的 check 是"检查"的意思。检查和检验的区别："检查"的意思是查看，重点在"查"，"检验"则是用一定标准予以验证，有时涉及测量，重点在"验"。英文 check 与中文"检查"的对应度比较高。但 check 也有"查核""核对"的意思，如 checklist（查核表），另见下例。

例 4.102：

原文：Check whether the data recorded fall into the range specified in the operation manual.

原译：检查所记录的数据是否在操作手册中规定的数值范围内。

改译：查核所记录的数据是否在操作手册中规定的数值范围内。

分析：本句原译将 check 译成"检查"也可以，但改成"查核"更准确，更专业，因为原文隐含着"核对"的意思。

例 4.103：

原文：As a result of scientific and technical progress it is possible to detect the presence of residues of veterinary medicinal products in foodstuffs at ever lower levels.

原译：由于科学技术的发展，现在能够探测出的饲料中存在的兽药产品残留物水平比以往任何时候都低。

改译：由于科学技术的发展，现在能够检测出的饲料中存在的兽药产品残留物水平比以往任何时候都低。

分析：本句中的 detect 应译成"检测"而不是探测。英文 detect 与中文"检测"的对应度比较高。

例 4.104：

原文：The Contractor shall also familiarize himself with the requirements of the other Trades, and examine relevant drawings and specifications.

译文：承包商应熟悉其他行业的要求，并检查相关图纸和规范。

分析：examine/examination 一般译成"检查"，意思是仔细检查。Check 与 examine 的区别：check 是检查某对象的性质或参数等是不是如预期一样，而 examine 则是仔细检查，看某对象是不是有什么异常，或者查看异常的原因。比如：check the engine 是指检查电机，确保其正常工作，而 examine the engine 则是仔细检查电机，看看电机为什么出现故障了，或者了解一下电机的工作原理。

但是，在自然科学领域里，examine 通常是"验证""观察""研究"之类的意思，具体视上下文语境而定。如：

例 4.105：

原文：To examine the reversibility of the drug effect on virus growth, the culture period was prolonged up to 16h p.i. after removal of (+)-catecin.

译文：为验证病毒生长的药物效应，将培养期延长至去除 (+)- 儿茶酸注射后 16 小时。

分析：本例的 examine 是指研究人员对某种现象或结论的确认，译成"验证"。另外，h.p.i. = hours post injection（注射后小时数）。

例 4.106：

原文：Unlike with our original potential, the new combined potential is sensitive to the difference between regular heavy baryons and heavy pentaquarks. We can examine whether a bound pentaquark state is possible within the combined potential by flipping the sign of the coupling constant of the baryon current term.

译文：与原电势不同，新总电势对常见重重子与重五夸克之间的差异敏感。我们可以通过重子流项的耦合常数的符号，验证总电势内是否可能存在束缚五夸克态。

分析：本例的 examine 与上例意思基本相同，也译成"验证"。

第五组词：

project, engineering, work/works

例 4.107：

原文：三峡水电站是世界上规模最大的水电站，也是中国有史以来建设最大型的<u>工程项目</u>。

译文：Three-gorges Hydropower Station is the largest hydropower station in the work and the greatest <u>project</u> ever completed by China.

例 4.108：

原文：在高速铁路向中西部延伸的过程中，沿线地质条件愈来愈复杂，<u>工程难度</u>不断增加。

译文：The geological conditions along the high-speed railway becoming more and more complicated, and the <u>engineering difficulties</u> keep increasing in Middle and West China.

例 4.109：

原文：北京首都国际机场 3 号航站楼的<u>工程</u>耗时三年多。

译文：The <u>works</u> of building the 3rd terminal of Beijing Capital Airport took more than three years.

综合分析：以上几例都出现了"工程"一词，但意思各不相同，以下是对"工程"的意思的小结：

单词	意思	说明
project	（工程）项目	project 指的是项目
engineering	工程（技术）	engineering 指的是工程中的技术设计等
work	（施工等）工程	work 指的是工程中的具体工作

第六组词：

oil，fuel，gas，lubricating oil

例 4.110：

原文：用新的石油勘探方法在沙漠里找油。

译文：A new prospecting method is adopted to search for oil.

例 4.111：

原文：飞机油的性能要求比汽车油高很多。

译文：Aviation fuel has much higher performance than car fuel.

例 4.112：

原文：汽车加油站

译文：Gas station

例 4.113：

原文：为保证磨床顺畅运转，需定期为其换油。

译文：Lubricating oil must be replaced regularly to guarantee the smooth operation of the grinder.

综合分析：以上几例都出现了"油"字，但译法各不相同。中文的"油"，可以译成 oil, fuel, gas, lubricating oil, grease, engine oil, edible oil 等英文。具体采用什么译法，必须根据上下文理解"油"的真正意思后才能确定。一般情况下，只要在翻译过程中具有准确区分"油"的含义的意识，就不会出现错误译法。

第七组词

customer，client

例 4.114：

原文：In 2007, Kampmann (Beijing) Co. Ltd. was founded to produce a range of products for heating, cooling and ventilation on the Chinese market and offer Chinese customers the complete service.

译文：卡普曼（北京）有限公司成立于 2007 年，可为中国市场提供一系列供暖、制冷、通风产品，为中国顾客提供一整套服务。

分析：customer 通常译为"顾客"，指购买产品或服务的人。

例 4.115：

原文：Bank client confidentiality means that the banks have a duty to keep confidential all facts that involve their clients.

译文：银行客户保密指的是银行对涉及客户的所有事实的保密义务。

分析：本句里的 client 意思为"客户"，指的是购买具有一定专业技术要求的产品的人，如银行、软件、律师、广告等服务类行业。

第八组词

loop，circuit

例 4.116：

原文：（输电线）接入系统方案以一回 230kV 架空线送出。

原译：(Transmission line) 230 kV overhead lines are used in the access system for power outgoing

改译：(Transmission line) one-loop 230 kV overhead lines are used in the access system for power out

分析：这里的"回"，是指回路。在输电线路的应用中，为了提高容量和可靠性，采用多回路并列运行，这个回路简称为"回"。

一回　one-loop/ 1-loop

二回　two-loop / 2-loop

以此类推。

例 4.117：

原文：It is difficult to make these circuit voltage controllable over a large frequency range.

译文：很难将这些电路的电压控制在一个较大的频率范围内。

分析：circuit 一般是"电路""回路"的意思。当 circuit 作"回路"解时，其意思也是电路。

第九组词

各种"机""器""床"等设备的翻译

中文的设备名称，通常都以"机""器""床"结尾。有些设备会有其他中文通名后缀，如"刀""子"等。对于这些设备名称的翻译，很多新译员都不多加思考，总爱把"机""器"就译成 machine，把"床"译成 bed，把"组"译成 set，不仅显得啰唆、不专业，有时候还是错误译法。以下是对这些词的译法的列表小结：

译法	举例
专用词	发电机—generator；切割机—cutter；焊机—welder；压机—press；打印机—printer；电梯/升降机—lift/lifter/elevator；钻机—rig/drill rig；变压器—transformer；曝气器—aerator；震击器—jar；显示器—display；离合器—clutch；驱动器—drive/driver；铣床—mill/miller；磨床—grinder；大剪刀—shear；刀具—tool；刀头—tool bit；刀架—tool rest；触子—contactor；反应堆—reactor；冰箱/冷藏库—refrigerator；发动机组—generator unit；泵组—pump unit
特殊译法	泥气分离器—poor boy degasser/mud-gas separator；采油树—Christmas tree
惯用组合译法	内燃机—internal combustion engine；空气压缩机—air compressor；气割机—gas cutting machine；倒角机—chamfering machine；真空吸尘器—vacuum cleaner；触电保护器—electric shock protector；滚床—roller machine；剪床—shearing machine；高炉—blast furnace；无轨电车—trolley bus；电葫芦—electric hoist
自行组合译法	咖啡机—coffee machine；自动埋弧焊机—automatic arc-submerging welder；履带式柴油起重机—crawler diesel crane
原则：一般能用专用词、特殊译法或惯用组合的，就不自己组合翻译。自己组合翻译时，注意词序要合乎英文表达习惯。总体而言，设备名词的译法，在准确的前提下，以简洁为宜。		

引申：

以上都是设备名称的译法，但实际上，其他类别的中文实物名称的英译，总体上也与以上设备名词的译法类似。

反过来，对于英文实物名称的汉译，总体上也可以遵守与以上设备名称中译英的规则相逆反的规则。比如，machine 不一定译成"机"；而 -er、-or 等后缀的名称，则可以根据实际情况译成"××机""××器""××者""××人"等。

三、"意外"词义

这里所讲的"意外"词义，其实是指某些词的不常见意思（熟词僻义）或者一些不常见的英文和数字组合，其中多数是第一次看到的人无法理解的。如：

family/range/series

例 4.118：

原文：When testing these amps with an APx585 Family analyzer, the AUX-0100 Switching Amplifier Measurement Filter should always be used, as it contains coupling capacitors to block the DC voltage from reaching the analyzer.

On the other hand, the APx525 Family, the 2700 Series, and the ATS-2 analyzers all have internal coupling capacitors that can be selected by choosing AC Input Coupling in the control or measurement software.

译文：用 APx585 系列分析仪测试放大器时，须连同 AUX-0100 转换放大器测量滤波器（AUX-0100 Switching Amplifier Measurement Filter）一起使用，因其包含耦合电容器，可以阻止直流电压到达分析仪。

另一方面，APx585 系列、2700 系列音频分析仪，以及 ATS-2 分析仪均有内置耦合电容器，这些电容器可以通过对控制软件或测量软件中的交流电输入耦合进行选择。

分析：本例中 Family 意思为"系列"。类似的表示系列的词还有 series、range，如：a series of static images（一系列静态画面），a range of ports（一系列端口，一些端口）。

termination

例 4.119：

原文：Dimensioned outline drawings show the plans and elevations of all equipment and components on the glycol contactor and regeneration package and location of piping/electrical/instrumentation terminations on the contactor and the package

译文：标明尺寸的外形图显示所有设备、乙二醇接触器和再生装置组件的平面图和高程，以及接触器和整套装置上的管道/电气/仪表终端的位置。

分析：本例中 termination 的意思是"终端"，电气终端（electrical termination）有时也称为终端器或终端电路（terminal resistor），是在传输线末端加上和传输线特性阻抗匹配的设备，以避免信号在传输线末端的反射。若传输线末端无终端设备，会让讯息失真，在数字系统上会造成不明确的数字信号准位，并让数位设备无法正常动作。在模拟信号系统上会产生鬼影,或是影响无线电发射器传输线上的功率边界。

correct

例 4.120:

原文:The copper resistance of each core of the completed circuits shall be measured and the value <u>corrected to an ambient temperature of 20℃</u>.

原译:应在 20℃环境温度下测量闭合电路各芯线的铜电阻,并调整电阻值。

改译:应测量闭合电路各芯线的铜电阻,<u>并将测量值修正至环境温度 20℃时的电阻值</u>。

分析:correct 这个词,译成"调整"似乎没什么不对。但从逻辑上考虑,电阻值测量后,一般是没必要调整的。有相关知识面的译员,应该知道,导体的电阻值会随着温度变化而变化。所以,在测量电阻值时,如果测量温度不是 20℃,就必须把实测电阻值用公式转化成 20℃时的电阻值,以便对不同温度下的电阻进行比较。这样的公式转化过程,就叫做修正。

对于 correct 的用法,与本例类似的例子还有:

例 4.121:

原文:The flow rate determined in the test shall be <u>corrected</u> to that applicable to a temperature of 30℃ using published data on variation in viscosity of water with temperature.

译文:试验中所测定的流速,应根据已发布的水粘度随温度变化的数据,<u>修正至适用于 30℃的流速</u>。

establish

例 4.122:

原文:In all cases where the service temperature is reduced by localised cryogenic storage or other cooling conditions, such factors shall be taken into account in <u>establishing</u> the minimum design temperatures.

译文:对于使用温度因局部低温储存或其他冷却条件而降低的所有情况,在<u>确定</u>最低设计温度时,均应考虑这些因素。

分析:本句中的 establish 是"确定",而不是"建立"。

major

例 4.123：

原文：The operator can hugely impact the system, making changes to extend the life of the controlled equipment by years, translating into major maintenance savings.

原译：操作员能够极大地影响系统，可进行变更使控制设备的使用寿命延长数年，这可转化成主要维护费用的节省。

改译：操作员能够极大地影响系统，可对系统做出变更，使其所控制设备的使用寿命延长达数年之久，从而节省大量的维护费用。

分析：本句中的 major 指的应该是"重大""大量"，而不是"主要"。

switch

例 4.124：

原文：As with the Ethernet switches, you can view the running configuration in order to do some sanity checking of the configuration on the terminal servers, if required.

译文：同处理以太网交换机一样，你可以查看运行配置，以便在终端服务器上对配置进行完备性检查（如有需要）。

分析：switch 意为"开关"，在网络设备中是指"交换机"。它可以为接入交换机的任意两个网络节点提供独享的电信号通路。最常见的交换机是以太网交换机。其他常见的还有电话语音交换机、光纤交换机等。

staff

例 4.125：

原文：The designed computer program allows expressing the digital image pixel values of the leveling staff code scales with numeric values.

译文：所设计的电脑程序可用数字值表达水准尺代码刻度的数字图像像素值。

分析：leveling staff 的意思是"水准尺""水准标尺"，其中 staff 是"标尺"。水准尺是水准测量使用的标尺，用优质木材或玻璃钢、铝合金等材料制成。

四、英文常见符号和包含数字的词

科技应用类文件的英文中，有些时候会为了简洁起见，使用一些缩写词。这些缩写词，有些在其他语境下也经常使用，但也经常会被用错。有些则是字母、符号、数字等几种形式的混合缩写。以下所列是相对常见的缩写词：

1. 易混淆或难记的缩写词

e.g. = for example：例如 i.e. = id est：即	etc. = et cetera：等等 et al = et alii：等人 et alibi：等地方
am = ante meridiem：上午 pm = post meridiem：下午	c/c 或 c.c. = carbon copy = being copied to：抄送 c/o = care of：转交 attn. 或 atten. = attention：收件人
ditto、di.、ibidem：同上 simili/same below：下同	A.D. = anno domini：公元 B.C. = Before Christ：公元前
S.N. = serial number：序列号；编号 No. 或 Nr. = number：编号；数量	p.a. = per annum：每年

2. 字母、符号、数字混合的缩写词

w/ = with w/o = without w/i = within	
7*24、7×24 或 24/7： 每周七天，每天 24 小时	VS = versus：……对…… 与……相对抗
Q1 = the 1st quarter：第 1 季度 Q2 = the 2nd quarter：第 2 季度 Q3 = the 3rd quarter：第 3 季度 Q4 = the 4th quarter：第 4 季度	M1 = the 1st month：第 1 个月 M2 = the 2nd month：第 2 个月 M3 = the 3rd month：第 3 个月 ……
W1 = the 1st week：第 1 周 W2 = the 2nd week：第 2 周 W3 = the 3rd week：第 3 周 ……	D1/ Day1 = the 1st day：第 1 天 D2/ Day2 = the 2nd day：第 2 天 D3/ Day3 = the 3rd day：第 3 天 ……

本节课后练习

请复习本节各部分内容。

第五章　疑难点解析

在应用类文件中，经常会碰到一些疑难点无法解决。其中有些疑难点，很明显就可以看出其难点，比如句子结构复杂，不好理解；或者句子中某个重要的词或词组无法理解，但却难以通过词典或网络确定其意思，从而导致句子意思不明。另外还有一些疑难点，看似不难，实则暗藏玄机，里面有意料之外的知识点，或者译员在翻译时，隐约会觉得译文不妥。其中前者即明显的疑难点，如果译员态度足够认真，相对比较容易解决，或者至少会意识到难点，并请教有经验的人帮忙解决。而后一种，即不明显的疑难点，则不仅需要译员具有足够的语言敏感性和严谨的翻译态度，还需要有宽广的知识面和熟练的解决疑难点的技巧。

在介绍疑难点解析方法时，必须说明的一点是，在翻译过程中，尤其是在理解难点，确定词义或句意时，如果通过词典或网络进行相关搜索，或者使用逻辑推理等方法，找到两个或更多个可能的解释，则与原文语境或背景最吻合（即最佳关联），通常就是原文真正想表达的意思。这应该也是关联理论的一种解释。

翻译过程中碰到的疑难点多种多样，本书无法一一予以说明解决方法。但是，由于缩写词占了疑难词的相当大一部分，下文将先介绍英文缩写词的缩写规则，然后介绍英文缩写词的理解与翻译方法，最后再介绍更具体的疑难点解决方法，即查词法、句子理解、回译、原文纠错等方面，列举一些典型例子，对疑难点的各种解决方法做一定的阐述。

第一节　英文缩略词的缩略规则[①]

本节简要地分类总结了英文缩略词的缩略规则，并且列举了较多的例子。目的一是让大家能够对各类缩略规则有更直观的认识，二是让大家值此学习机会，快速掌握某意思。以便将来碰到陌生的缩略词时，更轻松地运用学习的规则和联想法，更快速地查到缩略词的真正意思。

一、一般缩略规则

（一）首字母缩写（Acronym、Initialism）

1. Acronym 是由各单词的首字母（或前几个字母）组成的缩写。Acronym 这种缩写必须作为一个单词来发音，这一点经常被人们所忽略。例如：

纯首字母缩写：

AIDS [eɪdz] = acquired immune deficiency syndrome 艾滋病（获得性免疫缺乏综合征）

ASCII [ˈæskɪ] = American Standard Code for Information Interchange 美国信息交换标准码

BIOS [ˈbaɪəʊs] = Basic Input/ Output System （电脑）基本输入输出系统

ELISA [ɪˈlaɪzə, ɪˈlaɪsə] = Enzyme-linked immunosorbent assay 酶联免疫吸附测定

LASER [ˈleɪzər] = light amplification by stimulated emission of radiation 激光（受激辐射光放大）

NATO [ˈneɪtəʊ] = North Atlantic Treaty Organization 北大西洋公约组织

RADAR [ˈreɪdɑː(r)] = radio detection and ranging 雷达（无线电探测与测距）

STEM [stem; stɛm] = science, technology, engineering and mathematics 科学技术工程与数学领域

SONAR [ˈsəʊnɑː(r)] = sound navigation and ranging 声纳（音频导航与测距）

VIES [vaɪ] = VAT Information Exchange System 欧盟增值税信息交换系统

[①] 本部分内容参考了维基百科关于 Abbreviation、Acronym 和 Initialism 的介绍。

音节或音节首字母缩写：

Benelux [ˈbenɪˌlʌks]: a politico-economic union of three neighbouring states in western Europe: Belgium, the Netherlands, and Luxembourg 比荷卢经济联盟

Gestapo [gesˈtɑ:pəu] = Geheime Staatspolizei 盖世太保（德国秘密警察）

Interpol [ˈintəpɔl] = International Criminal Police Organization 国际刑警组织

UNIVAC [ˈju:nivæk] = Universal Automatic Computer 通用自动计算机

注意：

需要指出的是，有些缩略词忽略源短语的虚词或非重要单词，找不到某些词的代表字母，有些缩略词的字母则涵盖源短语包括虚词在内的所有单词。

2. Initialism 也是由一些名词或短语的字母（通常是首字母）组成。但与 Acronym 不同的是，Initialism 这种缩写不作为一个单词发音，而是一个个字母念出。例如：

ASAP = as soon as possible 尽快

CPU = central processing unit 中央处理器

FBI = Federal Bureau of Investigation （美国）联邦调查局

GDP = gross domestic product 国内生产总值

GMP = good manufacturing practice 良好生产惯例（国际通行的药品生产和管理规范）

GMT = Greenwich Mean Time 格林尼治标准时间

GPS = global positioning system 全球定位系统

HTML = hypertext markup language 超文本标记语言

MTBF = mean time between failures 平均故障间隔时间

NDT = non-destructive test 无损检验

OCA = oculocutaneous albinism 眼皮肤白化病

PVC = polyvinyl chloride 聚氯乙烯[①]

[①] 化学、生物、药学等行业中，有大量由多个单词组合而成的复合词，这些词通常都是名词术语，这些复合词在缩写时，通常都必须取好几个核心字根的首字母。如：acute myelomonoblastic leukemia（AMMOL，急性髓单核细胞性白血病）、adenosine triphosphate（ATP，三磷酸腺苷）、ethyl cyanoacetate（ECYA，氰乙酸乙酯）、polytetrafluroethylene（PTFE，聚四氟乙烯）、polyurethane（PU，聚氨酯）、small nuclear ribonucleo proteins（snRNP，小核核糖蛋白）。这种情况在其他行业也不少见，如：hardware（HW，硬件）。

REACH = Regulation for Registration, Evaluation, Authorisation and Restriction of Chemicals（欧洲）化学品注册、评估、授权与限制条例

SN = serial number 序号，系列号

UAV = unmanned aerial vehicle 无人机

YTD = year to date：年初至今

3. 根据读者或上下文而定，既可视为 Acronym，又可视为 Initialism 的缩略词：

FAQ ([fæk] or ef-ei-kju:) = frequently asked question 常见问题

IRA (['aɪrə] or ai-a:-ei) = individual retirement account 个人退休金账户

SQL (['si:kwəl] or es-kju:-el) = Structured Query Language 结构化查询语言

SUV ([sʌv] or es-ju:- 'vi:) = sports utility vehicle 运动型多用途车，多用途跑车

4. 一部分为 Acronym，另一部分为 Initialism 的缩略词：

CD-ROM [si:-di:- 'rɒm] = compact disc read-only memory 只读光盘存储器

CMOS [si:'mɒs] = complementary metal oxide semiconductor 金属氧化物半导体

IUPAC [ai-ju:- 'pæk] = International Union of Pure and Applied Chemistry 国际纯粹与应用化学联合会

MS-DOS [ˌem-es-'dɒs] = Microsoft disk operating system：微软磁盘操作系统

JPEG ['dʒeɪ-pɛg] = joint photographic experts group：联合图像专家组（一种数字影像压缩标准）

（二）截短缩写（Clipping）

1. 截头（initial-clipping）

copter = helicopter 直升机

chute = parachute 降落伞

phone = telephone 电话

plane = aerophane 飞机

pop = popular music 流行音乐

pub = public house 酒吧，客栈

roach = cockroach 蟑螂

2. 截尾（final-clipping）

ad = advertisement 广告

exam = examination 考试，考查

fax = facsimile 传真

gas = gasoline 汽油

gym = gymnastics / gymnasium 体操 / 健身操，健身房

info = information 信息

memo = memorandum 备忘录

photo = photograph 照片

stereo = stereophonic 立体声

sub = submarine 潜艇

taxi = taxicab 出租车

3. 截中（medial-clipping）

Dr. = doctor 医生，博士

expt = experiment 实验

maths = mathematics 数学

specs = spectacles 眼镜

ft = foot 英尺

BK = bank 银行

Qz = quartz 石英

RGE = range 量程

Qty = quantity 数量

WT = weight 重量

HRS = hours 小时（数）

4. 截头尾 / 留中（front-backclipping）

flu = influenza 流感

frig / fridge = refrigerator 冰箱

jams / jammies = pajamas / pyjamas 睡衣

5. 拼接（blending）

Chinglish = Chinese English 中式英语

motel = motorist hotel 汽车旅馆

smog = smoke + fog 烟雾

telex = teleprinter + exchange 电传

cablegram = cable telegram 海底电报

HiFi = high fidelity 高保真

NODE = noise diode 噪声二极管、二极管

Sci-Fi = science fiction 科学幻想

WiFi = Wireless Fidelity 无线保真（局域网）

6. 以辅音字母开头的词，取全部或部分辅音字母

TRML = terminal 终端

Mfr= manufacturer 制造商，厂商

LNR = linear 线性

7. 以元音字母开头的词，取首字母加全部或部分辅音字母

Asgd = assigned 指定的

Engr = engineer 工程师；Engrg = engineering 工程

EM = electromagnetic 电磁（的）

EM = electromechanical 电机（的）

注意：

地名不得缩写，但地名里的有些通用名可以取其关键字母缩写。如：

To 收件方 / 收件方地址

Fm. = from 发件方 / 发件方地址

Add. = Address 地址

Attn. / Atten. = Attention 收件人

c.c. = carbon copy = being copied to 抄送

c/o = care of 转交

RE. = reply 回复

Ave = Avenue 大街

Bldg. = Building 楼，幢

Blvd = Boulevard 大道

Fl = Floor 楼，层

Rd = Road 路

Rm = Room 房

Sq. = Square 广场

St. = Street 街

= No. 号

二、其他缩略规则

（一）利用数字、读音或意思缩略

B2B = business to business：企业间电子商务，企业对企业，商家对商家

i2L = integrated injection logic：集成注入逻辑

Y2K = Year 2000：二〇〇〇年

G20 = Group of 20：二十国集团

LC50 = 50% lethal concentration：（毒物）半致死浓度

PM2.5 = particulate matter 2.5：直径小于或等于 2.5 微米的颗粒物 / 2.5 微米以下颗粒物

$(ISC)^2$ / (ISC-squared) = International Information Systems Security Certification Consortium：国际信息系统安全认证协会

3M = Minnesota Mining and Manufacturing Company：3M 公司（明尼苏达矿业和制造公司）

4GL = Fourth Generation Language：第四代语言

i18n = internationalization：国际化（其中的 18，是指 i 和最后一个 n 之间共有 18 个字母）

X-conn = cross connection：交叉连接

Xmas = Christmas：圣诞节

XL = extra large：特大码

μp = microprocessor：微处理器

IOU = I owe you：欠条

VA：视在功率（原意为视在功率单位：伏安）

Var：无功功率（原意为无功功率单位：乏）

（二）利用符号或代号缩略

DRHO = delta rho = $\triangle \rho$：密度差

Delta time = $\varDelta T$：时间差

R&D = research and development：研发

B&F = beverage and food：饮料和食品

L/C = letter of credit：信用证

D/A = document against acceptance：承兑交单

O-ring：O 型圈

w/ = with

w/o = without

w/i = within

7*24 / 7×24 或 24/7 = 24 hours a day, seven days a week; all the time：全天候（即每周七天每天 24 小时）

（三）缩写套缩写

BREF = BAT Reference Notes= best available technique Reference Notes：最佳可行技术参考文件

LDIF = LDAP directory interchange format = lightweight directory access protocol directory interchange format：轻量级目录访问协议目录交换格式

VIES [vaɪ] = VAT Information Exchange System：欧盟增值税信息交换系统

（四）假性缩写

BTW = by the way：顺便提一下

CU = see you 或 CUL = see you later：回头见，再见

HRU = how are you

IOU = I owe you：欠条

IC = I see：知道了

LOL = laughing out loud：哈哈……

Q8 = Kuwait：科威特

（五）有些缩略词的最后一个词不仅在缩略词里有体现，还再次作为单词跟在缩略词后

ATM machine = automated teller machine：自动柜员机

HIV virus = human immunodeficiency virus：人体免疫缺陷病毒（即艾滋病毒 HIV）

LCD display = liquid crystal display：液晶显示器

PIN number = personal identification number：个人身份代码

（六）英文中收录的源自其他语种的缩略词

A.M. (*ante meridiem*) = before noon：上午

P.M. (*post meridiem*) = after noon：下午，午后

A.D. (*Anno Domini*) = in the year of our Lord：公元
（英文 B.C. (Before Christ)：公元前）

q.d. (quā que diē) = every day：（用药）每天一次

b.i.d. (bis in diē) = twice a day：（用药）每天两次

a.c. (ante cibum) / p.c. (post cibum)：饭前/饭后

ditto / di., dem / id., ibidem：同上

e.g. = exempli gratia = for example：例如，如，举例

i.e. = id est = that is to say, in other words：即，换句话说

etc. = et cetera [it'setrə] = and so on：（事或物）等，等等

et al = et alii：以及其他人

et al = et alibi：以及其他地方

（七）参数、量级与计量单位的缩写

1. 参数的缩写：

ln = Napierian Logarithm：自然对数，纳皮尔对数

V_i = instantaneous velocity：即时速度

D_i / D_o = inside diameter / outside diameter：内径/外径

DN20 = nominal diameter 20mm：公称直径 20mm

Max / Min = maximum / minimum：最大值/最小值

2. 量级的缩写

f = femto：飞（10^{-15}）

fs（飞秒）

p = pico：皮（10^{-12}）

pF（皮法）

n = nano：纳（10^{-9}）

nm（纳米）、ng（纳克）、ns（纳秒）

μ = micro：微（10^{-6}）

μm（微米）、μg（微克）、

m = milli：毫（10^{-3}）

mm（毫米）、mg（毫克）、ms（毫秒）

c = centi：厘（10^{-2}）

cm（厘米）

d = deci：分（10^{-1}）

dm（分米）、dB（分贝）

k = kilo：千（10^3）

km（千米/公里）、kg（千克/公斤）、kJ（千焦）

M = million：兆（10^6）

MW（兆瓦）、MΩ（兆欧）、Mbps（兆字节/秒）

G = gillion：吉（10^9）

GW（吉瓦）、Gbps（吉字节/秒）

3. 计量单位的字母大小写规则：

（1）人名转化过来的单个单位，首字母必须大写，非人名转化过来的单位，除 L（升）外，首字母都不大写。如：

七个基本物理量

量	代号	单位符号	单位符号
长度	l	米（公尺）	m
质量	m	千克（公斤）	kg
时间	t	秒	s
电流	I	安［培］	A
热力学温度	T	开［尔文］	K
物质的量	n	摩［尔］	mol
发光强度	I_v	坎［德拉］	cd

其他单位举例

ft（英尺）、in（英寸）、m^3（立方米）、g（克）、N（牛）、min（分）、h（时）、V（伏）、W（瓦）、Ω（欧）、Hz（赫）、℃（摄氏度）

（2）带有量级或复合单位，其中的每个组成单位，分别按单独的一个缩略词进行大小写。所以，有些单位首字母不大写，但中间或后面却有些字母要大写，有些全大写，有些全小写，但正确写法就是这样。如：

km（千米）、m/s（米/秒）、m/s^2（米/秒^2）、ms（毫秒）、Btu（英制热量单位）、oz（盎司）、bbl（桶）、MMcfd（百万英尺^3/天）、MW（兆瓦）、

kJ（千焦）、kWh（千瓦时）、mA（毫安）、pF（皮法）、dB（分贝）、mph（英里/小时）、rpm（转/分）、Ω·m（欧·米）。

（3）注意：

（a）有些字母大小写表示不同意思，如：

K：开尔文（温度单位）—k：千（量级单位）

M：兆（量级单位）—m：米（长度单位）/毫（量级单位）

S：西门子（电导单位）—s：秒（时间单位）

（b）个别字母可以表示不同的单位，如：

2g：2克，2倍重力加速度（地球上重力加速度为 9.81m/s^2）

（c）计量单位与货币单位单复数同形，如：

20kg、3A、40m、5.5Pa、45min、8in、100mph、USD300、Eur100。

三、缩略词的大小写

（一）缩略词里的小写字母通常是非核心词或虚词的缩写：

DoB = date of birth：出生日期

GoF = Gang of Four：四人帮

HBsAg = Hepatitis B surface antigen：乙肝表面抗原

LotR = Lord of the Ring：（电影）指环王

RoHS = Restriction of Hazardous Substances：（欧盟）有害物质限制指令

SoC = system-on-chip：晶片系统

u-POP = unexpected persistent organic pollutant：意外的持久性有机污染物

vPvB = very persistent very bioaccumulative (substance)：高持久性高累积毒性（物质）

（二）但是，并不是代表非核心词或者虚词的字母就必须小写，有些也直接大写，有些则省略掉，如：

ASTM = American Society for Testing Material：美国试验材料协会

ROA = return on assets：资产收益率

SWIFT = Society for Worldwide Interbank Financial Telecommunications：环球银行金融电信协会

VOA = Voice of America：美国之音

WIO = water in oil：油包水

WOB = weight of bit：钻压

四、名词缩写的复数形式

对于名词的缩写，通常可把缩略语视为普通可数名词。因此，缩略语和缩略词的复数形式，一般直接在其后面加小写的 s。例如：

Radars：雷达

BOPs = blow-out preventers：防喷器

BOQ = bills of quantities：工程量清单

L/Cs = letters of credit：信用证

CPUs = central processing units：中央处理器

HRSGs = heat recovery steam generators：余热锅炉

VIPs = very important persons：贵宾

本节课后练习

1. 写出以下缩略词的全称，理解其缩略词规则并说明哪些词有时可以不译或通常不译：

APPX	DWG	IT	Q1/Q2/Q3/Q4
ASME	EDTA	KFC	SCADA
ASTM	EPC	LOD50	SCUBA
B/L	EU	MODEM	SGR
BBC	FCPA	NAFTA	SOHO
CEO	FIFA	NASA	SOP
CET-6	FW/HW/SW	NCR	THAAD
CFS/CY	Hi-tech	Nd:YVO4	TOEFL
CNOOC	IATA-DRG	OEM	T/T
Co., Ltd.	IBM	O&M	UEL/LEL
Dept	IELTS	OHSAS	UNESCO
DIN	ILU	pcs	VAT
DIY	IOC	pH	WEEE
DNA/RNA	I/O	ppm	WWII
DNV	ISO	QA / QC	WWW

2. 试着找出缩略词为两个或更多个代表核心词义的字母的复合词，如：software（SW）、polytetrafluroethylene（PTFE）。
3. 了解计量单位的缩略词中，哪些是人名转化过来的，哪些不是？并记住计量单位字母大小写规则。

第二节　查词法

这里所讲的查词法，主要是针对翻译中碰到的难以确定意思的缩写词、疑难词，乃至不理解的句子，通过联系上下文、词典、网络搜索、逻辑推理等方法确定意思并加以翻译的方法。下文将通过具体案例说明各种查词法。但是，为了加强学习效果，请大家同时按照案例所述方法，做相应的思考和搜索。

一、英文查词

（一）英文缩写词

前两节介绍了缩写词的缩写规则及大体的理解和翻译方法。但是真正碰到缩写词的时候，经常还是无法立即理解，这时候，就只能通过词典和网络搜索解决了。

例 5.1：

原文：RoHS, items we purchase from You which are and should apply to RoHS standards should be marked with RoHS in some way on all boxes/bags or on the label.

原译：无铅。我方从贵司采购的条目应当并且应适用于无铅标准。因此，在所有箱子或袋子或标签上应贴上"无铅标志"。

改译：RoHS 标志：我方从贵司采购的条目应当适用于 RoHS 标准，所有箱子或袋子或标签上均应贴上 RoHS 标志。

分析：这里的 RoHS，在网上查，可以得知其意思是 restriction of hazardous substances，是欧盟限制使用有毒有害物质的禁令，该禁令共列举了六类有毒有害物质：铅（Pb）、镉（Cd）、汞（Hg）、六价铬（Cr6+）、多溴二苯醚（PBDE）、多溴联苯（PBB）。这是译员需了解的常识。

例 5.2：

原文：It should also be noted that, at welded piping joints where CS joins SS there is a stress concentration whenever there is a thermal transient event.

原译：应当指出的是，在 CS 与 SS 相连接的焊接管道接头处，只要出现瞬态热，就会出现应力集中。

改译：应当指出的是，在碳钢与不锈钢相连接的焊接管道接头处，只要出现瞬态热，就会出现应力集中。

分析：本句原译没有译出 CS 和 SS。其实只要在百度里搜索 "CS/SS 是什么材料"或用类似的关键词搜索一下，都可以得知其意思为"碳钢/不锈钢"。

这里想说明的一点就是，在翻译中碰到的英文缩写，只要不是行业内习惯使用的英文缩写，如 WTO，NBA 等，一般都要尽可能译出。

例 5.3：

原文：For the highest resolution AMOLED and LCD displays, the Prexision-80 gives unsurpassed photomask quality.

原译：对于分辨率最高的主动式有机发光显示器（AMOLED）和液晶显示器（LCD），Prexision-80 的光掩膜质量是无与伦比的。

改译：对于最高分辨率的有源矩阵有机发光二极管（AMOLED）和液晶显示器（LCD），Prexision-80 的光掩膜质量是无与伦比的。

分析：AMOLED = Active-matrix organic light emitting diode：有源矩阵有机发光二极管

active 有源；passive：无源

LCD = liquid crystal display：液晶显示器

这两个词的意思都可以直接从网络词典里查到。

例 5.4：

原文：Research clearly shows that D/M water is not safe drinking water.

原译：已有研究明确说明，D/M 水不是安全的饮用水。

改译：已有研究明确说明，去矿质水不是安全的饮用水。

分析：在 Bing 里查"D/M water"，可发现较多的 demineralized water，因此可以推测 D/M water 就是 demineralized water，即"去矿质水"。

典型缩写：

例 5.5：（借代式缩写）

原文：Rated Motor Tr Rated Motor Torque. The value entered into this parameter should come directly from the Motor Nameplate. If the nameplate only

lists the Motor Horsepower, the torque can be calculated from the equation:
Torque (ft-lb) = (HPx5252)/Rated rpm

原译：Rated Motor Torque（额定电机扭矩）。输入该参数的数值应直接来自电机铭牌。如果铭牌仅列出电机功率，则扭矩可通过下列方程式计算得出：

扭矩（ft-lb）=（功率 x5252）/ 额定转速

改译：Rated Motor Torque（额定电机扭矩）。输入该参数的数值应直接来自电机铭牌。如果铭牌仅列出电机马力数，则扭矩可通过下列方程式计算得出：

扭矩（ft-lb）=（马力数 x5252）/ 额定分钟转速数

分析：这里的 horsepower 意思，跟"功率"差不多，但实际又有所区别，指的是用马力表示的功率数值，这可以从 **Torque (ft-lb) = (HPx5252)/ Rated rpm** 这一公式中看出，公式中 HP 就是 horsepower，但由于涉及数字的计算，只能把 HP/ horsepower 理解成以马力表示的功率数，称为"马力数功率"或者"马力数"。同理，rpm 也要译成"分钟转速"或者"分钟转速数"。

本文件里类似的表达方法还有 rpm，secs 等，很多时候要把 rpm 译成"rpm 转速"，把 secs 译成"秒数"。类似的例子还有 kWh（千瓦时）、sqm（平方米）、psi（磅/平方英寸）、Btu（英热单位）等。

小结：
确定英文缩写词意思的几种主要方法：
（1）直接到上下文搜索；
（2）直接查网络词典；
（3）在网络上用辅助词 stands for、meaning、is 等，或者用中文辅助词"是""意思"等搜索；
（4）把数个缩写词一起搜索，即"捆绑搜索"；
（5）综合利用以上各种方法，并用逻辑推测确定辅助词。

（二）英文疑难词

例 5.6：

原文：The decimal expansion of a number may terminate (in which case the number is called a regular number or finite decimal, e.g., 1/2=0.5), eventually become periodic (in which case the number is called a repeating decimal, e.g., 1/3=0.333...), or continue infinitely without repeating (in which case the number is called irrational).

原译：数字的十进制展开式可能会终止（这时将该数称为正规数或有限小数，如：1/2 = 0.5），最终周期性循环（这时将该数称为循环小数，如：1/3 = 0.333...），或者持续无限不循环（这时将该数称为无理数）。

改译：数字的十进制展开式可能会终止（这时将该数称为正规数或有限小数，如：1/2 = 0.5），最终周期性循环（这时将该数称为循环小数，如：1/3 = 0.333...），或者持续无限不循环（这时将该数称为无理数）。

分析：decimal expansion：十进制展开（式）；repeating decimal：循环小数；irrational (number)：无理（数）。这里需要注意的是，句子中的 decimal 出现了三次，但有两个不同意思。

例 5.7：

原文：Move the cursor to highlight one of the rows (1 through 24) in this column and press ENTER.

The file name will begin with the day of the month (1 thru 31) and the current hour and minute.

原译：移动光标，高亮显示该栏中任一行（1 通过 24），然后按 ENTER（回车）键。

文件名将以日期（1 通过 31）和当前的时间开始。

改译：移动光标，高亮显示该栏中任一行（1 至 24 行），然后按回车键。

文件名将以日期（1 至 31 日）和当前的时间开始。

分析：在网络词典查 thru，其意思与 through 相同，所以可以推测 through 是 thru 的不规范缩写。但找不到符合本句语境的准确解释。不过，通过第二句的语境 day of the month (1 thru 31)，可以推测 thru = through，意思为"从……到/至……"。

例 5.8：

原文：Upon mutual written agreement between Buyer and IR, Buyer may cancel, reduce in quantity or reschedule Orders for HiRel Products as follows: Cancellation; Reduction in Quantity: Orders for HiRel Catalog Discrete Product (packaged and die form) may be reduced in quantity or cancelled (i) ninety (90) days prior to the initial estimated delivery date; or (ii) if IR has not delivered the Product by the initial estimated delivery date.

原译：经买方和国际整流器公司双方书面协议，买方可如下所述，将 HiRel 产品订单取消、减少订量或改期：（i）在最初的预计交货日期九十（90）天之前；或（ii）在国际整流器公司在最初的预计交货日期之前未交付产品的情况下，将 HiRel 目录个别产品（已包装，加模具形式）订单减少订量或取消。

改译：经买方和国际整流器公司双方书面协议，买方可如下所述，将 HiRel 产品订单取消、减少订量或改期：（i）在最初的预计交货日期九十（90）天之前；或（ii）在国际整流器公司在最初的预计交货日期之前未交付产品的情况下，将 HiRel 目录个别产品（已包装裸芯片形式）订单减少订量或取消。

分析：本段中 die form 这个词组的意思很难确定。如下图左侧的搜索记录所示，用 HiRel Catalog Discrete Product 或 Discrete Product 做辅助词查 die form，都得不到 die form 的合理解释。

但搜索"die form 产品"时的搜索结果有一条如下所示：

> 裸芯片_百度百科
> baike.baidu.com/view/1391960.htm
> 裸芯片（die，bare die，chip，die form，wafer form）半导体元器件制造完成，封装之前的产品形式，通常是大圆片形式（wafer form）或单颗芯片（die form）的形式存在，…

因此，可将 die form 译成"裸芯片"。

例 5.9：

原文：The life of a conventional coiled tubing string is normally limited by the fatigue life of its <u>bias weld</u>. As the required yield strength of coiled tubing increases, the fatigue life of the conventional <u>bias weld</u> decreases relative to the rest of the coiled tubing string. This limits the useful life of the entire string to the shorter life of the <u>bias weld</u>.

原译：传统连续油管柱的使用寿命通常受其<u>斜面焊缝</u>的抗疲劳寿命限制。随着对连续油管屈服强度的要求越来越高，传统<u>斜面焊缝</u>的抗疲劳寿命比连续油管柱的其余部分更短。这使整条油管柱的使用寿命<u>斜面焊缝</u>的更短的使用寿命所限制。

改译：传统连续油管柱的使用寿命通常受其<u>偏置焊缝</u>的抗疲劳寿命限制。随着对连续油管屈服强度的要求越来越高，传统<u>偏置焊缝</u>的抗疲劳寿命比连续油管柱的其余部分更短。这使整条油管柱的使用寿命被<u>偏置焊缝</u>的更短的使用寿命所限制。

分析：第三版《英汉石油技术词典》对 bias weld 的解释是"焊缝偏差"，这一解释让人感觉似乎有错。在网络上查 bias weld 不容易查到中文意思，但由于在查 bias weld 的同时经常会同时看到 butt weld，因此把两个词组同时查，可发现，bias weld 一般被译成"偏置焊缝"。

例 5.10：

原文：Turbo-alternator is equipped with resistance thermometers Pt of <u>100 ohms 0 □</u> for remote temperature measuring.

原译：远程温度测算使用配备有 <u>100 ohms 0</u> 的 Pt 电阻温度计的汽轮交流发电机。

改译：汽轮交流发电机配备有 <u>0℃ 100Ω</u> 的铂电阻温度计，用于远程温度测量使用。

分析：原译文里的"Pt of 100 ohms 0□"有点语焉不详的感觉，这里分析如下：
（1）其中 ohms 是原译文里没有译出来的部分，是电阻单位，可改

译成 Ω；

（2）Pt 这个词，原译文也没译出，其实这里应该译出，其意思是"铂"，原文的 resistance thermometers Pt 应该是"铂电阻温度计"的意思；

（3）0 后面的□，看起来很不起眼，很多人会忽略掉，但忽略掉之后，又不知道这个 0 是什么意思。在百度里以"电阻温度计 PT 100Ω 0"查找，部分结果如下：

基于Pt100 的电子温度表设计 - 期刊论文 - 道客巴巴
电子温度计的检测范围 ～为 - 40 ℃ 120 ℃。 关键词：单片机 铂电阻 ...Pt100 的电子温度表设计 43 Pt100 铂电阻的测温范围是 - 200 ℃ 600 ℃在 ...
www.doc88.com/p-30873066287.html 2011-1-21 - 百度快照

电阻温度计Pt 100 型号系列用于食品制药生化GA2540 设 - docin.co...
热度：浏览:25 评论:0 顶:0 收藏:0 举报文档介绍 电阻温度计Pt 100 型号系列用于食品制药生化GA2540 设 文档分类 生活时尚 -- 健康 文档标签： 螺纹(...
www.docin.com/p-34869452.html 2010-9-20 - 百度快照

铂电阻温度计-Pt100_电阻器_精诚亿想-中国领先的电子元器件分销商...
100-249CNY 0.85 制造商： 库存编号：1577819 商品型号：探头温度传感器 RPL... 感应器类型 铂电阻温度计-Pt100（0oC时100Ω） 主体的材料 316不锈钢 精度...
www.jc-ic.cn/show-1577819-1.html 2010-5-7 - 百度快照

铂电阻(Pt100)测量温度的范围是多少_百度知道
在检定各种标准温度计和精密温度计时作标准使用。基本参数 温度范围 0℃ 419.527℃ 东方晨景科技有限公司专业提供美国Lake Shore公司的铂电阻温度计，测量温度范围14 K...
zhidao.baidu.com/question/61842793.html 2009-8-28 - 百度快照

其中上图中的第三条有个"（0oC 时 100Ω）"的字眼，根据对相关铂电阻温度计知识的查找和了解（详见下图），知道铂温度计通常都是在 0℃时电阻为 100Ω，可以初步确定"□"其实是"℃"这个词在电脑里变成乱码了。

电阻温度计

利用导体电阻随温度变化而改变的性质而制成的测温装置。通常是把纯铂细丝绕在云母或陶瓷架上，防止铂丝在冷却收缩时产生过度的应变。在某些特殊情况里，可将金属丝绕在待测温度的物质上，或装入被测物质中。在测极低温的范围时，亦可将碳质小电阻或渗有砷的锗晶体，封入充满氦气的管中。将铂丝线圈接入惠斯通电桥的一条臂，另一条臂用一可变电阻与两个假负载电阻，来抵偿测量线圈的导线的温度效应。电阻将按下列公式随温度发生变化：

R=R0（1+aθ）

式中R是θ℃的电阻，R0是0℃时的电阻，a是常数。比较精确的式子是：

R=R0（l+aθ+bθ2）

所以，在经过查证后，将原文的译文修改为："远程温度测量使用配备有 0℃ 100Ω 的铂电阻温度计的汽轮交流发电机。"

本节课后练习一

翻译下列句子，注意下画线部分：

1. The 10-50 mA input signal wires are connected to the same pins in parallel with the resistor, with positive connected to J2-pin 12, negative to J2-pin 14.
2. The P&ID is particularly important for the development of start-up procedures when the plant is not under the influence of the installed process control systems.
3. The secondary winding provides a full load current of 30 A at a P.F. of 0.8.
4. As the heat dissipation rate is higher in ONAF transformer cooling system than ONAN cooling system, electrical power transformer can be put into more load without crossing the permissible temperature limits.
5. The main advantage of SVCs (Static VAr Compensators) over simple mechanically switched compensation schemes is their near-instantaneous response to changes in the system voltage.
6. Marine and salty environments also lower the lifetime of GI sheets because the high electrical conductivity of sea water increases the rate of corrosion, primarily through converting the solid zinc to soluble zinc chloride which simply washes away.
7. This system is capable of producing customer specified sizes in multiples of 600 x 600 mm at proportionally reduced finishing speed.
8. BMCR and TMCR are varied by increasing the value of steam flow rate of superheaters and reheaters.
9. Linear Dimensions - Lengths used to describe the nominal sizes of products should be expressed to the nearest centimeter (cm). Therefore, linear dimensions will be described as an integral number of centimeters (no decimals).
10. Frequent-flyer program is an airline promotional scheme designed to win the loyalty of frequent fliers.
11. ASME B16.5 - Pipe flanges and flanged fittings, NPS 1/2 through NPS 24
    ASME B16.47 - Large diameter steel flanges NPS 26 through NPS 60

12. The two most widely accepted methods for calculating Gauge **Repeatability** and **Reproducibility** are Average and Range Method and ANOVA (Analysis of Variance) Method.

二、中文查词

本部分介绍的是中文词（包括缩写词）的意思确定与翻译。①

例 5.11：

原文：工程涉及的设计及建安工程费用应由双方协商支付。

译文：The expenses for design and building installation work shall be paid by both parties through negotiation.

分析：本句的"建安工程"其实是中文缩写词。由于该词可能有多种意思，必须到网络上查找确定。在百度上查找"建安工程"，结果如下：

再查找"建筑安装工程"，可以知道其意思是建筑物或构筑物的安装工程，因此译成 building installation work。

---

① 本节网址查询日期均为 2020 年 6 月 2 日。

例 5.12：

原文：<u>正版药</u>价格很高，在于其高研发成本，更在于市场垄断。

原译：Reasons for high prices of <u>genuine medicine</u> include not only heavy R&D costs but also market monopoly.

改译：Reasons for high prices of <u>original medicine</u> include not only heavy R&D costs but also market monopoly.

分析：本句中翻译的难点在于"正版药"。关于其译法，可在网络上这样查找：

（1）先在百度里查"正版药"，得知其别名为"原研药"。

（2）用正版药、仿制药、原料药这三个名词，加上 medicine 一起在百度上搜索，也就是查"正版药、仿制药、原料药 medicine"，结果不理想。

（3）再查"原研药 medicine"，结果也不理想。

（4）再查"原研药 英文"，结果如下：

点击以上搜索结果的第一条链接，结果如下：

originator product 就是原研药，originator 指最早开发这个药的厂商，但这个称呼业内用得较少。常用的是 original drug/product，另外还有其他称呼：

innovator drug/ product——原研药都是新药，innovator 表示它的创新性；

brand name drug, branded drug——原研药都会申请商标（国内有区分商标和商品名，国外不区分）。

generic drug 是仿制药，"a generic equivalent" 这里的 equivalent 是指仿制药和原研药质量等同，生物等效。开发仿制药要先进行药学研究，证明药学上质量等同后再进行生物等效性试验（bioequivalence study，BE study），证明临床上生物等效。

a generic equivalent 就是这么来的。

对于称谓，业内各种叫法都有，中文也不是只有"原研药"一种称呼。药品审评中心（CDE）的老师有的爱说"原研药"，有的爱说"原发药"，我们原研药企业的人自己比较习惯说"原研药"，而较少说"品牌药"（尽管英文可以直译为"品牌药"）。现在很多外企在开发"品牌仿制药"（branded generics，branded generic drugs），为避免混淆，我个人喜欢用"原研药"和"original product"。

从以上内容的说明可以得知，回答这个问题的人，应该有一定专业水准，可以采用他给的"原研药"几种译法，但建议还是用 original drug 好一点，在特定语境下，可译成 original product。如果要避免在同一文章里反复使用同一词汇，则可变换用词，将 original drug、original product、branded drug、brand name drug 等换着用，但还是要以 original drug 为主。

例 5.13：

原文：矿石成分：P 比较高，块矿品位 52% 左右的含磷 0.5%，精矿粉中含磷 0.1%；含硫低，<u>双零</u>；含硅 6% 左右

译文：Ore composition: high P content: 0.5% P in lump ores with the grade of about 52%, and 0.1% P in concentrate fines; low sulphur: <u>below 1%</u>; silicon: about 6%

分析：这一句的重点是要弄清楚缩写词"双零"的意思。通过网络搜索"矿物含硫 双零 意思"可知，"双零"是指小数点后有两个零，如 0.002、0.003，因此译成 below 1%。

例 5.14：

原文：水泥混凝土路面

以水泥混凝土<u>面层</u>和<u>基（垫）层</u>组成的路面。

沥青混凝土路面

由适当比例的各种不同大小级配的矿料和沥青在一定温度下拌和成混合料经摊铺压实而成的<u>面层</u>和<u>基（垫）层</u>组成的路面。

……

沥青路面结构层可由面层、基层和垫层组成，层与层之间应紧密结合。

（a）面层应坚实、平整、耐磨。沥青面层尚应具有良好的防渗、抗滑、耐高、低温稳定等性能。面层厚度应符合表 12 的规定；

（b）基层应具有足够的强度、水稳定性和低温稳定性。表面必须平整，拱度应与面层一致，基层宽度应比面层每侧宽出 25cm。面层厚度应符合表 12 的规定。

译文：Concrete pavement

The pavement consisting of concrete surface and base (bed) course.

Asphalt concrete pavement

The pavement consisting of surface and base (bed) course formed by paving and compacting the mixture composed of mineral aggregates and asphalt graded in different sizes proportionately.

……

The asphalt pavement structure course may consist of surface course, base course and bed course, and the combination between different courses shall be close.

（a）The surface course shall be firm, flat and wear-resistant. In addition, the asphalt surface course should have good anti-seepage and anti-skid capability, good resistance to high temperature, and low temperature stability, etc. The surface course thickness shall conform to the provisions of Table 12;

（b）The base course shall have sufficient strength, water stability and low temperature stability. Its surface must be neat, whose camber shall be consistent with the surface course, and the base course width shall be higher than each side of the surface course by 25 cm. The base course thickness shall conform to the provisions of Table 12;

分析：在本例原文相距较远的段落中，出现了很多的"面层""基层"和"垫层"以及"结构层"和"层与层之间"等表达，而且译员在前面全文的定义中将"基层""垫层"译成 base course 和 bed course，将面层译成 surface，而在后面将所有"层"都译成 layer。总体上，让人感觉不大统一。

但是，敏感性比较高的译员，可能会发现，在网络词典里，"基层""垫层"都是译成 base course 和 bed course（不是 cushion layer），而且，"面层"也是译成 surface course（而不是 surface 或 surface layer）。

既然"基层""垫层"和"面层"的"层"都译成 course，是不是可以合理推定 course 就是"层"的意思，而且在本例中，所有"层"都要译成 course。带着这个问题，到网络词典里查 course，但海词在线、爱词霸、有道等网络词典里的单独一个 course 都没有"层"的意思。这时，大家是不是会想起收词比较全面比较专业的中国在线翻译网 http://chinafanyi.com/结果一查，还真有意想不到的收获：

- ⑧【建】(砖石砌的)一层，层；
a damp- course
防潮(湿)层.
lay the courses
砌砖.

原来，在建筑专业中，"层"就是要译成 course，这才是专业的译法。因此，原文中的所有"层"都要改译成 course，而不是用原来的 layer。列举本例的目的是为了提醒译员，在翻译时，对术语的翻译不仅必须做好前后一致性，还要注意从个别用词现象中推理出用词规律，这个规律，在本例中就是"层"的专业译法。

例 5.15：

原文：白垩化生屑灰岩：由岩溶作用形成的为高孔高渗储层，可看出最具有代表性的特征是"低电阻率"，由于钻井过程中泥浆液的侵入，侵入直径大于深浅侧向的探测半径，电阻率曲线为 Rs ≈ Rd < 1.0Ω.m；其他曲线特征类似于厚壳蛤生物灰岩，即"三低二高"的特征，低伽马（GR < 20API），低密度（RHOB < 2.5 g/cm³），高中子（NPHI > 15%），高声波（DT > 60μs/ft）。①

---

① 原文里说明是"三低二高"，但后面只列出了两低两高。根据上下文推测或者网络搜索，知道是少了"低阻"，即低电阻率。可能是作者认为前面已经说明过"电阻率曲线为 Rs ≈ Rd < 1.0Ω.m"，即电阻率较低，所以不再说明"低阻"。

译文：Chalky bioclastic limestone: the high-porosity and high-permeability reservoir developed under the effect of karstification is typically characterized by "low resistivity." Due to the invasion of mud during drilling, the invasion diameter is larger than the radius of detection in deep and shallow lateral direction, the resistivity curve is Rs≈Rd<1.0Ω.m; other curve characteristics are similar to those of rudistid biolithite limestone, namely, "three lows and two highs," including low gamma (GR<20API), low density (RHOB<2.5 g/cm^3), high neutron (NPHI>15%) and high interval transit time (DT>60μs/ft).

分析：在原文"高声波（DT > 60μs/ft）"里，括号里有个 DT，而 DT 是声波时差的代号，所以，可以知道这里的"高声波"是指"高声波时差"。另外，其他几个参数的说明如下：

英文缩写	意思及解释
GR	= gamma ray = gamma ray log[①]：自然伽马测井 自然伽马测井（GR）记录的是地层所含放射性元素衰变时放出的伽马射线，单位是 pulse/min（脉冲/分钟）或 API 单位，后者是美国石油协会（API）采用的单位，以两倍于北美泥岩平均放射性的模拟地层为刻度标准，其自然伽马测井曲线值的 1/200 定义为 1API 自然伽马测井单位，这种刻度单位已被各测井公司普遍采用。
RHOB	= rho bi (ρb) = bulk density[①]：体积密度（岩石的真密度）， 体积密度是单位体积中的岩石骨架和孔隙流体的质量总和。测定岩石密度的目的是通过密度进一步求得岩石的孔隙度。 另外，DRHO = delta rho = △ρ：用于补偿泥饼影响的校正值（泥饼影响校正值）。
RHOB	△ρ=(ρl-ρs)/k ρl 和 ρs 是根据长短源距计数率得到的，就是计数率构成脊肋图计算得出的，k 可能是经验常数。

---

① 这部分解释参考了网络链接 http://www.agoil.cn/bbs/simple/?t280970.html（2020 年 6 月 2 日检索）。

(续表)

英文缩写	意思及解释
RS, RD	RS = resistivity shallow（浅侧向电阻率） RD = resistivity deep（深侧向电阻率） 对于岩石电阻率和岩性、储层物性、含油性有密切的关系，因此通过研究岩石电阻率的差异可区分岩性、划分油水层、进行剖面对比等是电阻率测井的主要任务。
NPHI	= neutron porosity hydrogen index：中子孔隙度指数 中子测井孔隙度的单位是以水为 100 PU（孔隙度单位）来刻度的，其对应的"含氢指数"为 1。中子测井孔隙度可用于研究地层的孔隙度、岩性以及孔隙液体性质等地质问题。
DT	= delta time = △T：声波时差 声波时差是指接收声波的时间差值。利用这个差值可以进行相关运算，求解各种量值。

本节课后练习二

翻译下列句子，注意下画线部分：

1. 夹套上<u>多余的</u>连通管管口，应在系统吹扫合格后隔热施工前予以封焊。
2. SMT 贴片<u>生产线</u>优点是元器件安装密度高，易于实现自动化和提高生产效率，降低成本。
3. 方解石和白云石、韭闪石、透辉石、橄榄石等多数以<u>它形</u>不等粒变晶结构出现，而黄铁矿有时以自形半自形粒状变晶结构出现。
4. 环境友好型钻井液 (EFD) 用堵漏剂 CX-217，是天然植物高分子复合材料，本产品具有高黏度、高<u>静切力</u>的性能，其耐温性能达到 150℃以上，在饱和盐水中具有良好的<u>黏切性能</u>。
5. 目前国际上油菜品种改良的热点是将双高油菜品种改良为<u>双低油菜品种</u>。
6. 清洁作业区必须达到 30 万级以上净化车间的要求，具备独立的空调系统和空气净化系统。

三、结合逻辑与知识面

例 5.16：

原文：Cassette chamber　　　　　Dry pump

　　　(1) ultimate pressure: 10 Pa or better

　　　(2) Pumping time: within 10 min. from the atmospheric pressure to 20 Pa

原译：盒室　　　　　　　　　干式泵

　　　（1）极限压强：10 Pa 或 10 Pa 以上

　　　（2）抽运时间：10 分钟内从大气压强值增长到 20Pa

改译：盒式腔　　　　　　　　干式泵

　　　（1）极限压强：10 Pa 或更佳

　　　（2）抽气时间：10 分钟内从大气压强值抽到 20Pa

分析：本例的 or better 译成"以上"是不对的。这里的 or better 是指真空度更高，而真空度更高指的却是更低的压力，也就是说，比 10Pa 更低才叫 better。另外，对于第二小句，稍具物理常识的人都会知道，大气压远远比 20Pa 高，所以不能译成"增长"，而应该译成"抽到"。

例 5.17：

原文：Stator windings shall be connected in star.

　　　If no differential protection is specified, the star point may be concealed.

　　　In case current transformers have to be installed in the star point box, details will be given by Contractor.

原译：应对定子绕组进行星形连接。

　　　如无另外规定保护措施，应隐藏汇接点。

　　　若需将电流变压器安装在星点盒中，承包人应提供安装细节指导。

改译：定子绕组应为星形连接。

　　　如无另外规定差动保护措施，应隐藏中性点。

　　　若需将变流器安装在中性点接线盒中，承包人应提供安装细节指导。

分析：本句的 star、star point 和 star point box 是密切相关的一组词，其中 star point 可在网络词典里直接查到，其意思为"星点；中性点"，指的是变压器、发电机的绕组中与外部各接线端间电压绝对值相等的点。当电源侧（变压器或发电机）或者负载侧为星形接线时，三相线圈的首端（或尾端）连接在一起的共同接点称为中性点，简称中点。相应地，be connected in star 就是"星形连接"的意思，而由于 box 在电路连接中经常是"接线盒"的意思，所以 star point box 可译为"中性点接线盒"。

例 5.18：

原文：A pressure relief device or burst disc shall be provided, giving pressure release towards the motor in case of a short circuit in the box.

原译：应为接线盒配备泄压装置或爆破隔膜，在接线盒发生短路时，为电动机释放压力。

改译：应为接线盒配备泄压装置或爆破隔膜，在接线盒发生短路时，使压力向电动机内部释放。

分析：这里的 release towards the motor，字面意思为"向电机释放"，但译员却觉得这种译法好像不对，不敢这样译。

所以，这就涉及一个常识问题：对于电气电子器件，国际上有一种安全性要求，就是这些器件万一爆炸时，不能对人或者其他外物造成伤害，所以，在爆炸时，器件对外释放的碎片或其他有害物质不能太大，也不能超过一定距离，最好是直接爆向器件内部。

所以这里的 release towards the motor 还是应该译成"使压力向电动机内部释放"。

例 5.19：

原文：The HIV-1 and HIV-2 Positive Controls will produce a Reactive test result and have been manufactured to produce a very faint Test（"T"）line. The Negative Control will produce a non-reactive test result. Use of kit control reagents manufactured by any other source may not produce the required results, and therefore, will not meet the requirements for an adequate quality assurance program for OraQuicP ADVANCE Rapid HIV-1/2 Antibody Test.

原译：HIV-1 和 HIV-2 正控制会产生反应性试验结果，可生成非常微弱的实验线（T 线）。负控制则产生非反应性试验结果。任何其他厂商生产的试验盒控制试剂均不会产生要求的结果，因此不符合 OraQuicP

ADVANCE 的快速检测 HIV-1/2 抗体试验的适当质量保证项目的要求。

改译：HIV-1 和 HIV-2 阳性对照会产生反应性试验结果，可生成非常微弱的实验线（T 线）。阴性对照则产生非反应性试验结果。任何其他厂商生产的试验盒对照试剂均不会产生要求的结果，因此不符合 OraQuicP ADVANCE 的快速检测 HIV-1/2 抗体试验的适当质量保证项目的要求。

分析：在化学、生物、医药分析当中，control 通常是"对照"的意思，而 blank 通常是空白的意思，positive 和 negative 通常是阳性、阴性的意思。这里"对照"和"空白"的意思，可以到网络上搜索"阳性对照 阴性对照 空白对照 作用"。

比如，要检测某血清有没有感染某病毒，可以把已感染该病毒的血清作为阳性对照，未感染的作为阴性对照，待检血清的稀释液作为空白对照（背景对照），把这三组对照液与待检液，总共四组同时进行某性能的检测，然后将待检血清的检测结果与三组对照液的检测结果进行对照，以确定该血清到底是阴性还是阳性。

其中，阴性对照的目的是排除假阳性，阴性对照就是一个确定已知是阴性的东西，如果测试结果是这个样本显示阳性了，则说明你的实验有问题。

阳性对照的目的是排除假阴性，阳性对照就是一个确定已知是阳性的东西，如果测试结果是这个样本测不出阳性反应，则说明你的实验有问题。

空白有的时候可以当作是一种阴性对照。空白的特别意义在于测定实验的背景噪声。也就是说，空白样本被测得到的数值，告诉你未知样本的读数当中，有多少数值是背景，多少是真实特异性的反应所得。

例 5.20：

原文：The most common plug types are CEE yellow 2P+E, CEE blue 2P+E, CEE yellow 3P+E, CEE blue 3P+E and CEE red 3P+N+E. The colour of the casing refers to the regional electric power distribution at either 110/120 Volt = yellow, 230/240 Volt = blue or 400 Volt = red.

译文：最常用的插头类型为 CEE 黄色两相＋地线、CEE 蓝色两相＋地线、

CEE 黄色三相＋地线、CEE 蓝色三相＋地线，以及 CEE 红色三相＋零线＋地线。外壳颜色指的是地区电力配送，其中 110/120V 为黄色，230/240V 为蓝色，400V 为红色。

分析：本句原文中存在以下几个常识性难点：

（1）CEE = Conference on Electrical Engineering（电力工程会议）。

（2）2P+E、3P+E 和 3P+N+E 里的 P = phase(s)，N = neutral line（零线），E = earthing line（地线）。

（3）世界各国民用电压标准。世界各国民用用电所使用的电压大体有两种，分别为 100V~130V、220V~240V 两个类型。100V、110V~130V 被归类低压，如美国、日本以及全球大多数船上的电压。低压设计注重的是安全。220V~240V 则称为高压，其中包括中国的 220V 及英国的 230V，很多欧洲国家也在用 220V~230V 的电压。高压设计注重的是效率。一般说来同属一种电压体系的普通电器在各国间可通用，因为大部分的电器都有 20% 的电压浮动范围。

例 5.21：

原文：Vibration severity shall not exceed the values given in table 1 of IEC 60034-14 or GB 10068. Balancing and measurement shall be done with a half key fitted in the key way.

When performance guarantees have been requested in the requisition, tolerances shall be in accordance with table 8 of IEC 60034-1 or GB 755.

原译：振动烈度不得超过表 1 中 IEC 60034-14 或 GB 10068 所列各值。平衡调试及测量应在键槽的半键配合状态下完成。

若请购单对产品性能保障有所要求，则产品的允许误差应符合表 8 中与 IEC 60034-1 或 GB 755 的相关参数规定。

改译：振动强度不得超过 IEC 60034-14 或 GB 10068 的表 1 所列各值。平衡调试及测量应在键槽的半键配合状态下完成。

若请购单对产品性能保障有所要求，则产品的公差应符合与 IEC 60034-1 或 GB 755 的表 8 的规定。

分析：原文前后两小段里都有个"or"，乍一看，好像"or"后面的内容有点多余，不知道到底与 table 1 和 table 8 有无关联。但是，照字面意思来看，前一段的 table 1 of IEC 60034-14 or GB 10068，与后一段的"table 8 of IEC 60034-1 or GB 755"里的 or，应该分别是两个标准之间的连接词，也就是说 IEC 60034-14 和 GB 10068 是并列的，IEC 60034-1 和 GB 755 也是并列的，应分别译成"IEC 60034-14 或 GB 10068 的表 1"和"IEC 60034-1 或 GB 755 的表 8"。

但是，译员感觉这样的译法在逻辑上行不通。这里，希望大家了解一个知识点，国际标准化文件，包括 ISO（国际标准化组织）的标准文件和 IEC（国际电工技术委员会）的标准文件，有很多都是国际通用的标准，而且通常会被译成各国自己的语言，当作自己的国家标准；反过来，有时候，有些国家在某个行业因为处于全球领先地位或其他原因，其国家标准也会被翻译成英语（如果母语是英语的国家则不用翻译），当作 ISO 或者 IEC 标准，以作为国际通用的标准，然后各国再把这些标准翻译成自己的语言。当然，这些标准体系可能还不只是 ISO 或者 IEC，可能还包括其他一些行业协会的标准。

知道了这一点，也就不难理解本例的真正意思了。其实，IEC 60034-14 和 GB 10068 以及 IEC 60034-1 和 GB 755，分别是同一个标准的英文版和中文版，内容一样，所以才有原文这样的表达。大家可以通过网络查找对此予以确认。

例 5.22：

原文：All metal pipes for electrical cables shall be free from sharp edges, the ends of all pipes shall be filed smooth and <u>bends in pipes shall have a radius of not less than eight (8) times the outside diameter of the pipe</u>.

原译：电缆的所有金属管道应无锐边，所有管道两端应挫平，<u>管道内弯管直径最小为管道外部直径的八(8)分之一</u>。

改译：电缆的所有金属管道应无锐边，所有管道两端应挫平，<u>管道弯曲处半径至少应为管道外径的八（8）倍</u>。

分析：本句加下画线部分的难点，一是 bends 的意思，二是 a radius of not less than eight (8) times the outside diameter of the pipe 的语法和逻辑理解。

其中，bends 是 bend 的复数形式，通过网络词典可查到其意思为"弯曲""弯道"之类的意思，这里应理解为"弯曲处"。而 a radius of not less than eight (8) times the outside diameter of the pipe 则是"至少为管道外径八（8）倍的半径"。加下画线部分的意思，其实是指在布设管道时，管道的弯曲处的弯曲度不能过大，弯曲处的半径应至少为管道外径的八倍，否则管道容易被堵塞或损坏。

第三节　句子理解

一、关键词理解

例 5.23：（中文难句里的关键词）

原文：晋升条件不足时可设职务代理：

- 各级职务出现空缺时，若无具备晋升条件的人员派任，应提升适当人员代理职务
- 除任职年限不足外（以不足一年为限），其余条件不足者，不得提升
- <u>同等职位代理</u>，视代理期间工作绩效于适当时机办理直接调任；<u>不同职等代理</u>，<u>跨一职级</u>代理满半年，<u>跨两职级</u>代理满一年时，可办理晋升。

译文：Deputy can be created if the promotion conditions is insufficient:

- If nobody is qualified for the appointment when the position of any grade are vacant, suitable personnel shall be promoted as deputy.
- The employees whose seniority (less than one year) is insufficient, if any other conditions is also insufficient, shall not be promoted.
- The deputy for the position of the same grade can be directly transferred at proper time depending on the work performance of the deputy; the deputy for the position of different grade can be promoted if the acting period reaches half an year for the grade difference of one, or if the acting period reaches one year for the grade difference of two.

分析：本例里的"同等职位代理"的意思是"同等级职位的代理人"，"不同职等代理"是指"不同等级职位的代理人"。其中，"代理人"可译成 deputy。

例 5.24：

原文：Note, however, that such a relationship does not always predict a favorable effect, as illustrated by failure of drugs that effectively lower premature ventricular beat rates or raise high-density lipoprotein (HDL) cholesterol to have the expected cardiovascular benefits.

原译：但请注意，这一关系并非始终预示着有利的药物影响，正如药物无法做到有效降低室性早搏率或提升高密度脂蛋白（HDL）胆固醇，获得预期的心血管效益一样。

改译：但请注意，这一关系并非始终预示着有利的药物影响，正如可有效降低室性早搏率或提高高密度脂蛋白（HDL）胆固醇的药物却无法达到预期的心血管治疗效果一样。

分析：本例的原译，看起来好像没什么问题，实际上出现了重大理解错误。其中的关键点，是没看清楚 failure ... to 这一远距离搭配结构。原文 as illustrated by 后面的结构，实际上是 failure of drugs that ... to have the expected cardiovascular benefits，也就是说，that 是引导一个定语从句修饰 drugs，而不是 illustrated 的对象。

例 5.25：

原文：Developing CDWQGs

- Current published scientific research related to health effects, aesthetic effects, and operational and treatment issues are considered
- Health-based values are established on the basis of comprehensive review of the known health effects associated with the contaminant such as cancer, organ toxicity, life-stage effects
- Aesthetic effects (e.g., taste, odor) are taken into account when these play a role in determining whether consumers will consider the water drinkable
- Operational considerations are factored in when the presence of a substance may interfere with or impair a treatment process or technology (e.g., turbidity interfering with chlorination or UV disinfection) or adversely affect drinking water infrastructure (e.g., corrosion of pipes)

- Treatment technologies and analytical feasibility are also considered in setting the guideline value

原译：**加拿大饮用水水质指标的制定**

- 对有关健康效应、美学效应、操作和处理问题的目前已发表的科学研究加以考虑
- 以对关于癌症、器官毒性、生长阶段效应等污染物的已知健康效应的全面回顾为基础，建立以健康为基础的价值观
- 如果美学效应在决定消费者是否会考虑水可以饮用中起到作用，则对美学效应（如味道和气味）加以考虑
- 如果一种物质的存在可能会干扰或损害处理流程或技术，或对饮用水基础设施造成不利影响，则将操作事项归为考虑因素
- 确立指标价值时，还会考虑处理技术和分析可行性

改译：**加拿大饮用水质量指南的制定**

- 确立指标价值时，还会考虑处理技术和分析可行性
- 考虑与健康效应、感官效应、操作和处理问题相关的目前已发表的科学研究成果
- 基于与污染物有关的癌症、器官毒性、生长阶段效应等的已知健康效应的综合评审，建立健康价值观
- 如果消费者将感官效应（如味道或气味）作为判断水是否可以饮用方面的重要因素，则应考虑感官效应（如味道或气味）
- 如果一种物质的存在可能会干扰或损害处理流程或技术（例如，浊度影响氯化消毒或紫外光消毒），或对饮用水基础设施造成不利影响（例如，对管道造成污染），则应考虑操作事项
- 在设定指导值时，还应考虑处理技术和分析可行性

分析：（1）Health-based values：健康价值观

注意：像 Health-based values 这样的 A-based B，有时候可以省译 -based。

（2）第三小句中的 Aesthetic effects 里的 aesthetic 一词，不论是理解成"美学"还是"审美"都不恰当。但如果查中国译典，就可以发现 aesthetic 有"感官（的）"的意思，而这才是正确意思。

(3) determining 的意思，译成"确定""决定"都不妥。查中国译典，发现有"是……的决定因素"的意思，因此，可以认为 determining 是"在……方面起决定性作用"。同时结合上下文，即可改成改译后的译文。

例 5.26：

原文：

- In Ontario, all sources of water considered to be a precious resource and strong legislative mechanisms are in place to protect all waters for future generations
- Our Drinking Water Safety Net was designed to recognize that the protection of all water sources is the key to delivering safe, high quality water
- We continue to develop new standards for chemical contaminants and technology to ensure that high quality water continues to be delivered to the consumer

原译：

- 在安大略省，所有的水源均被视作宝贵的资源，且有力的法律机制已经就位，以为后代保护所有用水
- 我们饮用水安全网的设计目的是对把保护所有水源作为高质量安全供水的关键表示认可
- 我们仍在制定化学污染物的新标准和技术，以确保向消费者供应高质量用水

改译：

- 在安大略省，所有的水源均被视作宝贵的资源，且已建立有力的法律机制，以为后代保护所有用水资源
- 我们的饮用水安全网的设计目的，是让公民认识到，保护好所有水源是高质量安全供水的关键
- 我们将继续为化学污染物和相关技术制定新标准，以确保向消费者供应高质量用水

分析：本例原译第一小点表达不畅。第二、三小点的译文则过分机械，

其中第二点的 was designed to 之后相当于省略了两个词（比如 let citizens），所以宜将译文加以增译。第三点则没看清楚 new standards for chemical contaminants 是和 technology，都是 develop 的宾语。

例 5.27：

原文：Ground beef will be dark purple until it meets oxygen. That's why ground beef in the centre of a pack won't match its cherry-red surface colour.

原译：当碎牛肉遇到氧气时，颜色会变成深紫色。这就是一堆碎牛肉的中间部位颜色与表面的樱桃红色显得不太一致的原因。

改译：碎牛肉遇到氧气之前呈深紫色。这就是一堆碎牛肉酱的中间部位颜色与表面的樱桃红色显得不一致的原因。

分析：Ground beef will be dark purple until it meets oxygen 这一句原译文是"当碎牛肉遇到氧气时，颜色会变成深紫色"，乍一看似乎看不出问题。但在阅读英文原文时，可能会感觉 until it meets oxygen 这个意思不大确定，所以不敢确定原译文对还是不对。为了确认原文的意思，可以到网络上查看相关知识。经查阅后，可以知道牛肉遇到氧气前呈深紫色，遇氧气后变成红色。

经过这样的网络搜索之后，再回头查阅词典，可以发现，until 的意思是"直到……（时间）为止"，其英英解释还有"在……之前"的意思。所以，本例的 until 其实就是"在……之前"，并没有特殊的意思。

例 5.28：

原文：The metal used for bolts and screws should be stable and not subject to deterioration either naturally or after treatment. In particular, non-stainless steel items are protected by metal spray. This protection is carried out after machining on all threaded parts so they can be easily installed with no *play* or with very little *play* on tapped parts which are themselves protected.

译文：螺栓和螺钉采用的金属性质必须稳定，且自然状态下或经处理后不易变形。特别的是，非不锈钢物品应加以金属喷镀保护。所有螺纹零件均在加工后进行金属喷镀保护，以轻松地且使锥形部件无间隙或间隙极小地安装，这是其最好的自我保护。

分析：play 这个词，在本句中的意思很不好确定。到中国译典里查，能查到"自由活动（的余地）"，这个意思跟原文有点相关，怀疑是"间隙"的意思。在《新英汉科学技术词典》，可查到其名词解释第6条"（游，余，空，间，缝，齿）隙，间距，窜动量"，因此，可以确定play是"间隙"的意思。

例 5.29：

原文：Selenium, zinc, chromium, manganese and copper minerals are required for different enzyme functions. A deficiency or excess of minerals could be the cause of nutritional, toxic or metabolic disorders. The mere supplementation of micro-minerals in compound feed is not sufficient, the bioavailability of micro- minerals in the digestion tract of different animals should be taken into account as well as antinutritional factors, such as the phytic acid in cereals, which chelate mineral cations and proteins, forming insoluble complexes, which leads to decreased bioavailability of trace minerals.

译文：硒、锌、铬、镁和铜等矿物质是各种酶功能实现的必要元素。这些矿物质缺乏或过量都可引起营养失调、中毒或代谢紊乱。仅在复合饲料中添加微量矿物质是不够的，还应考虑到不同动物消化道中微量矿物质的生物利用度和抗营养因子，如谷类中的植酸可螯合矿物阳离子和蛋白质，形成不可溶配合物，最终导致痕量矿物质的生物利用度降低。

分析：本段译文中chelate mineral cations and proteins里的chelate是动词（"螯合"的意思），不是名词，其对象是cations and proteins，具体可以上网求证。

二、原文理解陷阱

（一）中文陷阱

例 5.30：

原文：蓄电池应避免阳光直接照射，远离火源，不能置于大量放射性、红外线辐射、有机溶剂和腐蚀气体环境中。

原译：Keep the batteries away from direct sunlight, fire source, a large number of radiation, infrared radiation, organic solvents and corrosive gases.

改译：Keep the batteries away from direct sunlight, fire source, intensive radiation, infrared radiation, organic solvents and corrosive gases.

分析："大量放射性"是典型的将形容词当作名词的错误中文表达，其意思实际上是"强辐射"，所以不译成 a large number of radiation，而是要译成 intensive radiation。

例 5.31：

原文：本书概述了历史上先后出现的主要区域文明及影响范围，认识其传统文化在现实生活中的表现。

原译：The book summarizes civilizations of main area and its scope of influence appeared successively in history, and get to know manifestation of its traditional culture in real life.

改译：The book summarizes major regional civilizations appeared successively in history and their scopes of influence, and understand the manifestations of their traditional cultures in real life.

分析："主要区域文明"的意思是"主要的区域性文明"而不是"主要区域的文明"。

本例中涉及的中文语法，是三个名词并列的"ABC"结构，其可能意思有："A 的 BC""AB 的 C""B 的 AC""A、B、C"等，具体需视上下文语境和中文习惯用法确定。

例 5.32：

原文：环境空气中，不含有腐蚀金属和破坏绝缘的有害气体或尘埃，使用时，不得使变压器受到水、雨、雪的侵蚀。

原译：The ambient air must not contain corrosive metal or harmful gas or dust which may cause damage to the insulation. When in use, the transformer must be prevented from erosion of water, rain and snow.

改译：The ambient air must not contain any harmful gas or dust which may corrode metals or damage the insulation. When in use, the transformer must be prevented from erosion of water, rain and snow.

分析：原文"腐蚀金属和破坏绝缘的有害气体或尘埃"的意思是"会导致金属被腐蚀或者导致绝缘被破坏的有害气体或尘埃"，而不是"腐蚀性金属和会破坏绝缘的有害气体或尘埃"。

例 5.33：（意思辨析）

原文：泵如长期停用，应将液缸体内的残留介质排空并冲洗干净，然后在泵体表面涂上防锈油，存放在干燥无腐蚀气体处，并加罩遮盖。

原译：If the pump will be out of service for a long time, the residual media in the hydraulic cylinder shall be drained and rinsed, then anti-rust oil shall be applied to the pump surface; the pump shall be stored in the environment with dry and non-corrosive gases and covered.

改译：If the pump will be out of service for a long time, the residual media in the liquid tank shall be drained and rinsed, with anti-rust oil coated to the pump surface; and the pump shall be stored in dry place without corrosive gases and covered.

分析：本句的"液缸"不是液压缸，而是存放液体的缸；"存放在干燥无腐蚀气体处"是指"储存在干燥而且无腐蚀气体的地方"，而不是"存放在具有干燥的非腐蚀性气体的地方"。本句更好的译法是：

The pump that will be out of service for long shall be stored in the place without corrosive gas and covered after draining and rinsing the residual media inside the hydraulic cylinder and coating anti-rust oil on the pump surface.

例 5.34：

原文：进货不合格品可采用"限定性合格"方式使用或退换货。

原译：Incoming nonconforming products can be used, returned or replaced in "restrictive conforming" manner.

改译：Incoming nonconforming products may be used in "restrictive conforming" manner, or returned or replaced.

分析：本句中的"可采用'限定性合格'方式使用或退换货"，是指采用"限定性合格"方式使用，如果不这样使用，就退货或换货。也就是说，"限定性合格"方式只是"使用"的方式状语，不是"退换货"的状语。

例 5.35：

原文：本机器主要用于花岗岩石材和大理石的柱座、柱帽自动仿型切割。如不同规格的圆柱、罗马柱等复杂形状。改变手工加工精度差、损耗大的缺点。

原译: The machine is mainly applied in the automatic shape cutting of the column bases of granite stone and marble and the column cap. The stone with a complicated shape like circular column of various specifications and Roman columns. To deal with the bad precision and great loss of hand finishing.

改译: The machine is mainly applied in the automatic shape cutting of granite and marble column bases and column caps, for example, complicated shapes such as circular columns and Roman columns of different sizes. It can overcome the problem of poor precision and great loss of manual processing.

分析: 本段问题点:

（1）原文第一小句有个小陷阱，即"柱座、柱帽"是并列成分，都受到"花岗岩石材和大理石"的修饰限定，但译员认为"花岗岩石材和大理石的柱座"与"柱帽"是相并列的。

本例中涉及的中文语法，是"A 和 B（的）C、D"，这种表达法，通常 A 不是独立的，而是 A 和 B 同时修饰 C 和 D。

（2）原文句子有点问题，第一、二个句号都用得不恰当，第三个分号应该是句号。本句相当于："本机器主要用于花岗岩和大理石的柱座、柱帽自动仿型切割，如不同规格的圆柱、罗马柱等复杂形状，可改变手工加工精度差、损耗大的缺点。"

例 5.36:

原文: 主营业务为工程咨询与工程承包，主要提供公路、桥梁、隧道、岩土、机电、市政、建筑、港口与航道等领域的勘察、设计、咨询、试验检测、监理、施工、总承包等工程技术服务。

原译: Its main businesses are engineering consultation and project contracting and providing professional service on highway, bridge, tunnel, ground, electromechanical engineering, municipal administration, building, survey, design, consult, test detection, supervision, construction, and general contracting of port and channel, etc.

改译: Main businesses: engineering consultation and project contracting, mainly providing engineering and technical services for survey, design, consultancy, test & detection, supervision, construction, and general contracting in highway, bridge, tunnel, geotechnic, electromechanical, municipal, building, port and channel engineering fields, etc.

分析：注意：本句所提供的"工程技术服务"的修饰语是"勘察、设计、咨询、试验检测、监理、施工、总承包等"，而"公路、桥梁、隧道、岩土、机电、市政、建筑、港口与航道等领域"则是修饰领域，不是与前者相并列的成分。

本例涉及的中文语法，是"A、B、C、D……等XX的E、F、G、H等YY"，是前两例语法现象的升级版。此类语法结构，通常都是"XX的YY"的意思，其中A、B、C、D等是XX的定语，E、F、G、H等是YY的定语，核心意思还是YY。

例5.37：

原文：井底破碎坑分布较规则，由若干个同心的破碎圆环组成，而每个破碎圆环则是由许多几何形状基本相似的单个牙齿形成的破碎坑组成。每个破碎坑的几何形状和尺寸，决定于载荷、侵入深度、牙齿的形状及其分布、加载时间、岩石性质等一系列因素。

原译：Bottomhole breaking pits are regularly distributed and consist of several concentric breaking rings, and each breaking ring consists of a breaking pit formed with many individual teeth that have a similar shape. The shape and size of each breaking pit are determined by loads, invasion depth, tooth shape and distribution, loading time period, rock properties, etc.

改译：Bottomhole breaking pits are regularly distributed and consist of several concentric breaking rings, each consisting of many breaking pits formed with individual tooth that have a similar shape. The shape and size of each breaking pit are determined by loads, invasion depth, tooth shape and distribution, loading time period, rock properties, etc.

分析：原文"而每个破碎圆环则是由许多几何形状基本相似的单个牙齿形成的破碎坑组成"里的"许多"指的是许多个"破碎坑"，而不是许多个"单个牙齿"，这从后一句中也可以证实。

例 5.38：

原文：肯尼亚分散式的电力输送系统也非常关键：<u>依靠进口昂贵的化石燃料进行大规模电力生产，维护任务繁重的电网和传输线路，均不适合散乱分布的村庄和偏远的城镇。</u>

原译：Distributed power transmission system in Kenya is also very critical: large-scale power production depending on fossil fuel imported expensively <u>to maintain</u> burdensome power grid and transmission line is not suitable for the scattered villages and remote towns.

改译一：Distributed power transmission system in Kenya is also very critical: <u>large-scale power generation</u> depending on expensive imported fossil fuel and <u>maintenance</u> of complicated power grid and transmission line with heavy maintenance load <u>are not suitable</u> for the scattered villages and remote towns.

改译二：Distributed power transmission system in Kenya is also very critical: <u>large-scale power generation</u> depending on expensive imported fossil fuel and power grid and transmission line with heavy maintenance load <u>are not suitable</u> for the scattered villages and remote towns.

分析：注意，这里的"维护……"不是前一小句"依靠……"的目的，原文的意思是：

依靠进口昂贵的化石燃料进行大规模电力生产以及维护任务繁重的电网和传输线路，都不适合散乱分布的村庄和偏远的城镇。

也就是说，"维护任务繁重的电网和传输线路"和前一小句"依靠进口昂贵的化石燃料进行大规模电力生产"，可以看作的并列主语，而"均不适合散乱分布的村庄和偏远的城镇"则是句子的谓语和宾语。

（二）英文陷阱

例 5.39：

原文：**Statistical analysis varies markedly from any OECD or other recognized approach, and does not establish toxicological relevance.** The authors used a multivariate technique called Partial Least Squares Discriminant Analysis (PLS-DA) to investigate the relationship among 48 blood and urine measurements relative to the different treatment groups. PLS-DA can be used to identify patterns in the data and to develop a function which can be used to discriminate between the groups. <u>However, just because you can discriminate between the groups does not make the result toxicologically relevant</u>. In fact, the majority of laboratory values in all groups appear to be within the normal range established by variation within the study. Fundamental data are either absent or, if present, are not presented in the typical format of mean values and standard deviations, which prevents the results from being evaluated appropriately.

原译：**经合组织或其他公认方法的统计分析结果差别很明显，且未建立毒理学相关性资料**。作者以被称为偏最小二乘判别分析法（PLS-DA）的二乘多变量法研究了与不同治疗组相关的 48 个血液与尿液测量结果之前的关联性，偏最小二乘判别分析法可用于确认数据模式并建立可用于判别不同治疗组的函数。<u>然而，正是由于可以判别不同的治疗组，使得结果不具有毒理学相关性</u>。实际上，所有治疗组的多数实验值看起来都在本研究所建立的正常变化范围内。基本数据要么缺失，要么就是未表现为平均值和标准差的一般形式，这就影响了结果的正常评估。

改译：**统计分析显著偏离任何一种经合组织或其他公认的实践的方法，并且未能建立毒理学的相关性**。作者以被称为偏最小二乘判别分析法（PLS-DA）的多变量法研究了与不同治疗组相关的 48 个血液与尿液测量结果之间的关联性。偏最小二乘判别分析法可用于确认数据模式并建立可用于区别不同治疗组的函数。然而，单靠能够区别不同的治疗组并不能证明结果具有毒理学相关性。实际上，这项研究中所有治疗组的多数实验值看起来都在它自己建立的正常变化范围内。基本数据要么缺失，要么就是这些数据未表现为平均值和标准差的正常形式，这就阻碍了对实验结果的正常评估。

分析：本例要说明的是原文中加下画线的句子的译法，即 <u>However, just because you can discriminate between the groups does not make the result toxicologically relevant</u>，这里的 just because，原译文理解成"正是因为"，

但实际上，just because 却与后面的 does not 相响应，意思却是"只因为……并不能"。这是一个很容易让人上当的语法陷阱。

例 5.40：

原文：**Data presented are highly sporadic and are not sufficient to support conclusions drawn.** While we understand that all data and all data analysis cannot be included in any publication of this nature, one would normally, for example, present data for both sexes rather than presenting data via one approach for male kidney outcomes and another for female kidney outcomes. This is important as it allows for determination of consistency of effect. Additional data sufficient to support the conclusions can easily be provided in supplemental online data tables.

原译：**所显示数据均很孤立，不足以支持所下结论。**虽然我们知道所有数据与数据分析均未在任何此类刊物发表，但你可以（比方说）显示两种性别的数据，而非通过一种途径只显示用于雄性动物肾脏结果，而通过另一途径只显示雌性动物肾脏结果。这一点很重要，因为这将直接决定效果的一致性。其他足以支持该结论的数据可通过补充的在线数据表轻易获得。

改译：**所显示数据均高度孤立散落，不足以支持所下结论。**虽然我们知道任何此类刊物都不可能涵盖所有的数据与数据分析，但通常人们会（比方说）显示两种性别的同类数据，而不是通过一种途径只显示用于雄性动物肾脏结果，另一途径只显示雌性动物肾脏结果。这一点很重要，因为这将直接决定效果的一致性。其他足以支持该结论的未发表的数据按理完全可通过补充的在线数据表轻易地提供给读者。

分析：本例与上例一样，都是容易让人上当的语法陷阱。本例中用到的 all ... not 是典型的部分否定句式，*all* data and *all* data analysis *cannot* be included in any publication of this nature 是指"任何此类刊物都不可能涵盖所有的数据与数据分析"，而不是"所有数据与数据分析均未在任何此类刊物发表"。

例 5.41：

原文：We believe the HCB and ANSES should help society understand this evidence rather than call for new long term tests for which there is no scientific need.

原译：我们相信，法国生态技术高级委员会和法国食品、环境与职业健康安全局与其帮助社会了解这些证据，不如呼吁进行没有科学要求的新的长期试验。

改译：我们相信，法国生态技术高级委员会和法国食品、环境与职业健康安全局与其呼吁进行没有科学必要性的新的长期试验，不如帮助社会了解这些证据。

分析：这里要注意 rather than 的用法，rather than 是指宁愿做 rather than 前面所述的事情，而不愿做 rather than 后面所述的事情。HCB = Haut Conseil des Biotechnologies（法国生态技术高级委员会），ANSES = Agence nationale de sécurité sanitaire de l'alimentation, de l'environnement et du travail（法国食品、环境与职业健康安全局）。

例 5.42：

原文：A broad range of scientists have strongly criticised the research on statistical grounds and because the strain of rats used are prone to develop cancer as they age anyway.

原译：许多科学家都基于统计学原因对该研究提出强烈批评，因为该品系大鼠在任何形式的衰老过程中产生癌症。

改译：许多科学家都基于统计学原因以及该品系大鼠会在任何形式的衰老过程中产生癌症的事实而对该研究提出强烈批评。

分析：原文的主句应该是 A broad range of scientists have strongly criticised the research，而后面的内容是两个理由。

例 5.43：

原文：The calculations of the environmental distribution and concentrations have been made using the EUSES 2 program, considering various combinations of the use patterns and **different rates** at **which** the substances and polymers may break down to PFOS in the environment.

The risk evaluation shows possible risks for secondary poisoning for all use areas in all of the scenarios used to examine the effects of different rates of break down and different combinations of releases.

原译：环境分布与浓度的计算以欧盟物质评估体系 EUSES 2 计划做出，并考虑不同使用模式与这些物质及其聚合物可能在降解时向环境释放的 PFOS 的不同比例的组合。

风险评估结果显示了所有使用区域在所有情况下都存在着二次中毒的风险，以用于检验不同降解率和不同排放组合的影响。

改译：环境分布与浓度的计算以欧盟物质评估体系 EUSES 2 计划做出，并考虑不同使用模式以及这些物质及其聚合物在环境里降解成 PFOS 的**不同降解率**的各种组合。

风险评估结果显示了所有使用区域在所有情况下都存在着二次中毒的风险，以用于检验不同降解率和不同排放组合的影响。

分析：本句难点在于 different rates 后面的部分。原译法让人感觉意思不明确，对原文理解不透彻。后来，在后文中又一次看到了 different rates，这次是以 different rates of break down 的形式出现，这是"不同降解率"的意思。由此可以推测，前面的 different rates 后面的 which 指的都是 rates，而 different rates 的意思就是以什么样的速率降解并向环境释放 PFOS。

在碰到复杂的意思不容易确定的定语从句时，要以不同的断句方式考虑原文的结构。

例 5.44：（后置定语或定语从句修饰范围）

原文：The procedures described in Sub-Clause 4.24.2 and Sub-Clause 4.24.4 shall prohibit the offering or providing, directly or indirectly, of anything of value, including cash, bribes or kickbacks to any employee, representative, customer or official in connection with any transaction or business dealing connected with the Project. This includes the prohibition to offer, promise or provide compensation for acceleration formalities.

原译：第 4.24.2 项条款及第 4.24.4 项条款中描述的行为规程应禁止向任何员工、代表、客户或与本项目相关的任何交易或交涉有关的高管直接或间接提供或给予任何有价之物，包括现金、贿赂或回扣。这包括禁止给予、承诺或提供用于加快手续完成的报酬。

改译：第 4.24.2 款及第 4.24.4 款所述规程应禁止向与本项目的任何交易或业务处理有关的任何员工、代表、客户或高管直接或间接提供或给予任何有价之物，包括现金、贿赂或回扣，包括禁止为加快相关手续而给予、承诺或提供报酬。

分析：原文加下画线有两处，原译都把后置定语的修饰范围弄错了。在这两处中，后置定语是修饰前面一串的词，而不是只修饰这个词。这是新译员常犯的错误。

例 5.45：

原文：Just because a job is deemed at risk from automation, it does not necessarily mean it will be replaced soon, notes Mr Frey.

译文：弗雷先生解释说："虽然某一工作存在被自动化取代的风险，但它不一定意味着不久就会被取代。"

分析：Just because ... it does not necessarily mean... 的意思是"仅仅因为……并不意味着……"，综合整个语境，参考译文译成"虽然……但它不一定意味着……"

例 5.46：

原文：These two inferences are no more contradictory than those two.

译文：这两个推论与那两个一样，都不会互相矛盾。

分析：no more ... than ...：和……一样都不；就跟……一样……

又如下例：

例 5.47：

原文：Elementary substance Calcium is not any more stable than Potassium, both do not exist naturally.

译文：单质钙跟钾一样都不稳定，都无法在自然界存在。

分析：not ... any more than：与……一样都不……语气比 no ... more than ... 更强。

例 5.48：

原文：State aid earmarked for specific sectors, or sectoral aid includes a number of measures targeting for instance: rescue and restructuring of firms in difficulty; Sectors covered include shipbuilding; steel; coal; land, sea and air transport; agriculture; fisheries and aquaculture.

原译：为某些特定部门提供的国家援助或跨部门援助包括各种措施，比如对困难企业进行援救和重组；涉及的部门包括造船、钢铁、煤矿、<u>土地、海洋和航空运输</u>、农业、渔业以及水产业。

改译：指定给特定部门的国家援助或跨部门援助包括各种措施，比如对困难企业进行援救和重组；涉及的部门包括造船、钢铁、煤矿、<u>陆海空运输</u>、农业、渔业以及水产业。

分析：从原文标点中看出，每个部门之间都用分号隔开，但 land, sea and air transport 这三个之间却用逗号隔开，在 transport 之后才是分号。而且"土地"和"海洋"跟其他词似乎不同类，所以，可以推测 land, sea and air transport 是一个整体，即"陆海空运输"行业。

本节课后练习一

翻译下列句子，注意下画线部分：

1. 薄膜类太阳能组件由玻璃、不锈钢、塑料、陶瓷衬底或薄膜上<u>沉积</u>的几微米或几十微米厚的半导体膜构成。
2. 为了提高设备的使用寿命,当接触器工作在关闭"Off"或故障"Failure"的时候，<u>必须关闭锁 SCR 门触发信号</u>。
3. <u>这种红外线橱柜感应灯</u>采用进口技术 MCU 电路设计而成，具有稳定好，抗干扰强等特点。
4. <u>空调</u>、通风系统垂直风管与水平风管连接<u>处</u>均设置 70℃的防火阀，风管穿越防火隔墙、楼板和防火墙处的孔隙用柔性防火材料填实封堵。
5. Just because something is legal doesn't mean it's the right way to do things.
6. There is also the Desert Rose whose trunk looks like nothing so much as a blooming elephant leg.
7. In 2000, 3M (a major global producer of PFOS based in the United States) announced that the company would phase out the production of PFOS voluntarily from 2001 onwards. <u>At a meeting of the OECD Task Force on Existing Chemicals following this announcement</u>, several OECD countries agreed to work together informally to collect information on the effects of PFOS on the environment and on human health to allow hazard assessment to be produced. This hazard assessment concluded that the presence and persistence of PFOS in the environment, as well

as its toxicity and bioaccumulation potential, indicate a cause for concern for the environment and for human health.

三、习惯性不规范表达

例 5.49：

原文：化学品存储地点要张贴 MSDS，并且根据 MSDS 设置相关个人 PPE。

译文：MSDS shall be posted at storage points of chemicals, and PPE shall be provided as required by MSDS.

分析：本句中文里面夹杂了两个英文缩写 MSDS 和 PPE。其中，MSDS = material safety data sheet，译文可以照抄这个缩写。但"个人 PPE"的表达却暗含陷阱，因为 PPE 是 personal protection equipment 的缩写，意思就是"个人防护用品"（较少说成"个人防护设备"），所以在英译时，不能译成 personal PPE，只能译成 PPE，即省略"个人"。

例 5.50：

原文：开关柜柜门上设有紧急分闸装置，为用电安全提供了绝对的保障。

原译：our switchgear boasts an emergency tripping device on the door, which provides absolute guarantee for electrical safety.

改译：the switchgear has an emergency tripping device on the door, which absolutely guarantee the electrical safety.

分析："为用电安全提供了绝对的保障"是很普通的一个中文句式，如果将其直译成 provides absolute guarantee for electrical safety，就很生硬。本句的"为用电安全提供了绝对的保障"可转化成"绝对保障了用电安全"，再译成 absolutely guarantee the electrical safety。

例 5.51：

原文：公司主要生产经营各类塑料编织制品和辐照高分子材料，拥有国际先进水平的塑料宽幅编织复合生产线和辐照生产设备，有一批高中级技术、管理人才及积累了多年实践经验的高素质员工。

译文：The company specializes in various plastic woven products and radiation resistant polymer materials. It boasts internationally advanced wide-weaving plastic composite production line and radiation equipment, along with middle-and-high-level technical and management talents and experienced employees.

分析：原文"有一批高中级技术、管理人才及积累了多年实践经验的高素质员工"译成 along with middle-and-high-level technical and management talents and experienced employees，其中"积累了多年实践经验的高素质员工。"是比较啰唆的中文表达，其实际意思就是想说明员工有经验或水平高，可直接简化翻译成 experienced employees。这是一种较好的处理办法。

例 5.52：

原文：内部分隔可将电弧的破坏性降低到最低程度。

原译：The inner separation can reduce arc destructiveness to minimum extent.

改译：The inner partition can minimize arc destruction.

分析："降低到最低程度"这个中式表达可以直接译成 minimize，这是最简洁的译法，应该也是最好的译法。其他类似的中文表达也可以这样处理。

例 5.53：

原文：测量由测量班对路面中心线及边线的位置和高程进行复测。

原译：The measurement is remeasured by the measurement class at the location and elevation of the center line and the side line of the road.

改译：It is remeasured by the measurement group at the location and elevation of the center line and the side line of the road.

分析：本句原文里出现了三个"测"字，感觉上有点啰唆。所以，译文不宜把三个"测"都直译出来，而是根据语境进行翻译。

四、填空式句子

（一）英文不完整句子

例 5.54：

原文：The next 10 years are absolutely crucial: Emissions will have to be on a sharp downward path by _____ for any hope of success. Greenhouse gases must be cut nearly in half from _____ levels.

原译：其中，未来十年至关重要：要想实现任何程度的成功，都必须用_____让温室气体排放量迅速降低，从_____水平接近减半。

改译：其中，未来十年至关重要：要想实现任何程度的成功，都必须在_____年前让温室气体排放量迅速降低，在_____年的水平上接近减半。

分析：本例是填空题，原文有两个空格，在翻译的时候（尤其是使用 Trados 时）不大好处理。碰到这样的原文时，译员不仅要注意必须把空格放在原文中合适的位置，还要知道空格里该填写什么性质内容。本例空格里填写的内容应该都是年份（其实分别是 2030 和 2010），所以，两个空格应分别改译为"在_____年前"和"在_____年的水平上"。

例 5.55：

原文：（以下是关于诺基亚手机的说明文件）：

Fill in the missing names of each key to complete the sentence below:
Data will be lost in keys [14] _____, [15] _____ and [16] _____ in a series _____ device and keys [59] _____, [60] _____ and [61] _____ in a series _____ device if the entry at key [34] _____ in a series _____ device and entry at key [117] _____ in a series _____ device are replaced.

原译：请填入所缺密匙的名称，补充下列句子：
在下列密匙中，数据会丢失：
诺基亚系列手机，密匙 [14] _____ [15] _____ 和 [16] _____ 以及诺基亚_____系列手机，密匙 [59] _____ [60] _____ 和 [61] _____ 诺基亚_____系列，若密匙 [34] _____ 条目被取代，[117] _____ 以及在诺基亚_____系列，若密匙 [117] _____ 条目被取代，则数据也会丢失。

改译：请在以下空格里填入密钥的名称，以将下列句子补充完整：
如诺基亚_____系列手机的密钥 [34] _____ 的条

目被取代，或诺基亚_____系列手机的密钥 [117] _____
_____的条目被取代，则诺基亚_____系列手机的密钥 [14]
_____, [15] _____和 [16] _____以及诺
基亚_____系列手机的密钥 [59] _____, [60] ____
_____和 [61] _____的数据会丢失。

分析：本例其实是考试中的一个完形填空，但这种中间带空格待填写的句子，也时不时地会出现在英译中文件里，有时也会出现在中译英文件里。由于我们的母语不是英语，本来做英译中翻译就不熟悉的人，碰到这种中间带空格待填写的句子，就更加找不着北了。对于本例，我们可以试着把句子整理成以下格式：

Fill in the missing names of each key to complete the sentence below:

Data will be lost in keys [14] _____, [15] _____ [16] _____ in a series _____ device and keys [59] _____, [60] _____ and [61] _____ in a series _____ device if the entry at key [34] _____ in a series _____ device and entry at key [117] _____ in a series _____ device are replaced.

把句子转化成这样的格式之后，感觉难度有所降低了吧。如果还不懂，再试着把所有空格和空格前的数字都删除掉，并把替换成 AA，BB，CC 等，结果如下：

Fill in the missing names of each key to complete the sentence below:

Data will be lost in keys AA, BB and CC in a series DD device and keys EE, FF and GG in a series HH device if the entry at key II in a series JJ device and entry at key KK in a series LL device.

这样转化之后，句子结构应该很清楚了吧。其意思是不是就如以上修改后的译文？

以后再碰到类似的中间带空格的句子时，就可以考虑用这种办法进行处理了，或者把空格当作一个句子成分，设想句子增加了这样的句子成分后应该怎么翻译，这样就可以看清楚句子结构，可以顺利地把句子译出来。不过，还要格外注意像本例中把一句分成好多行，甚至看起来像是好多段一样的情况，不要被看起来很长的样子迷惑了。

（二）中文不完整句子

例 5.56：

原文：针对此次喷漆室＿＿＿＿＿万立方米/小时废气和烘干室＿＿＿＿＿立方米/小时的废气设备的招标，投标方需要提供一种方案

原译：In terms of bid invitation for waste gas of spray booth (ten thousand m^3/h) and waste gas treatment equipment of drying room (m^3/h), the bidder should propose a scheme

改译：For the bid invitation for ＿＿＿＿＿ m^3/h waste gas treatment equipment of spray booth and ＿＿＿＿＿ m^3/h waste gas treatment equipment of drying room, the bidder should propose a scheme

分析：本句中的横线空格，是原文中要手工填写的地方，在翻译的时候不能漏译。而且，译文里的空格，还应放在合适的位置处。

例 5.57：

原文：×××时间，×××公路或地点，在×××对面或附近发生火工品运输车×××。

原译：A blasting supplies transportation vehicle * * * is found *** at the opposite to (or near) * * * highway (or site), at * * *(time).

改译：A blasting supplies transportation vehicle is found *** at the opposite to (or near) * * *, * * * highway (or other site), at * * *(time).

分析：注意，原文有四个空白处待填写，译文也有四处空白处，看起来好像没问题，但实际上漏了一个细节。

原译 opposite to (or near) 后的空白，对应于"在×××对面或附近"，而"×××公路或地点"前所留的空白，却被忽视了，跟 opposite to (or near) 后面的空白共用了。

但原译的 is found 后面却多了个空白，可能译员认为原文"运输车"后面的空格是对事件性质的说明，而原译文认为 vehicle 后面的×××是车牌号，所以，在 vehicle 后面加了空白，而这是错误的理解。所以，按该译员的思路，应该在 highway 前再加个空白，即译文应该有五个空白。

这就是翻译和校对中对细节的要求。

五、复杂句式理解

1. 长句

中译英长句

例 5.58：

原文：在管束间设计有合理的汽侧通道，以使其在包括最低循环水温等各种运行条件下有较佳的汽流分配，以减低汽阻损失和保证凝结水有<u>小过冷度</u>。

译文：Reasonable steam side channel is designed among pipe bundles, to ensure good steam distribution under all working conditions including the lowest temperature of circulating water, to reduce steam resistance loss and guarantee <u>slight condensate depression of condensed water</u>.

分析：本句的主句是第一小句，后两小句都是目的状语从句，而且第一个状语从句较为复杂。翻译中文复杂句时，不管是主句还是从句，都要先找出主干，再弄清楚其他成分与主干之间的关系。而且，通常情况下，可以先译出主干，再把其他成分逐步添加到合适位置，组成一个完整句子。译完整个句子后，还要检查有无漏译或者需修改、调整的地方，做好相应补漏和修改、调整。

例 5.59：

原文：<u>按引进技术设计制造的设备，在按引进技术标准设计制造的同时，还必须满足最新版的电力行业（包括原水电部、原能源部）相应规范</u>，当两者有矛盾时，以电力行业标准为准。

译文：<u>The equipment designed and manufactured with imported technical standards must also</u> conform to the latest power industry specifications (of original Ministry of Water Conservancy and Electric Power and original Ministry of Energy etc.), when conflict between them occurs, the power industry standards shall prevail.

分析：本例属于低复杂度的复杂句，共包含两个独立句子。第一句包括前三小句，主干是"设备（在……同时）还必须满足相应规范"，第二句是后两小句，主干是最后一小句。

翻译的难点是第一句的时间状语，即第二小句"在按引进技术标准设计制造的同时"。由于原文"按引进技术标准设计制造"与前面的"按引进技术设计制造的"所指内容相同，在英文里不宜重复译出，因此，直接把"按引进技术设计制造的设备"理解成"按引进技术设计标准制造的设备"，当作译文的后置定语，并省译后面的相同成分。

例 5.60：

原文：设备检修结束后，由设备主管工程师负责，会同修理单位技术负责人及工序使用人，按照合同附件中有关的检修任务书内容和验收标准共同进行初步验收，验收合格后方可交付工序投入使用，并进入设备保修期。

译文：After overhaul of equipment, the engineers in charge of the equipment, together with the technical director and process users, shall be responsible for the primary acceptance of equipment in accordance with relevant overhaul order and acceptance standards in the attachment of the contract. The equipment can be put into operation in the process and enter equipment warranty period only after passing the acceptance.

分析：本句是典型的中文复杂句。本句的结构分析如下：

（1）句子的主干是"设备工程师负责初步验收"。

（2）"设备检修结束后"是时间状语，"会同……按照……和验收标准共同"是方式状语。

（3）"验收合格后方可交付工序投入使用，并进入设备保修期"这一句的主语是"设备"，而不是"工序"。

例 5.61：

原文：为加强公司安全生产规章制度的管理工作，建立和完善公司安全生产规章制度管理体系，使安全生产规章制度的制定、修改、废止、审核、发布和备案工作程序化、规范化，提高工作效率，保证工作质量，促进依法治企，依据国家相关法律法规，制定本制度。

译文：This system is hereby formulated in accordance with relevant laws and rules of China to reinforce the management for safety production rules and regulations of the company; establish and perfect the management system of safety production rules and regulations of the company; routinize and standardize the formulation, revision, abolishment, review, publication and filing work of safety production rules and regulations; improve working efficiency; guarantee working quality; and expedite the governance of the enterprise by law.

分析：本句是程序文件里常见的介绍性句子，是句式结构相对简单的复杂句。像这种句子，除了英文选词要准确外，最重要的就是整个句子结构的选择。本句原文先把制定制度的目的说出来，最后才是主句，由主句说明依据相关法律法规，制定本制度。这是中文的习惯性表达。在英译时，不宜采用原文句式，而是应该把主句先译出，再说明制定本制度的依据，最后才把一长串的目的译出。只有这样，句子才会协调，才不会显得头重脚轻。

例 5.62：

原文：预计在未来几年内，提高创新能力、促进经济发展方式的转变、保护环境实现可持续发展将成为宏观经济发展的重要课题。<u>作为新兴产业之首的水务环保企业的行业发展目标顺应了目前国家宏观经济发展方向，将迎来更好的投资发展期</u>，同时企业也应有效树立改革理念，以适应快速变化的环境，同时把握经济改革中所带来的新机遇。

原译：It is expected that in the coming years, improving innovation capability, promoting the transformation of economic development mode, and protecting the environment to achieve sustainable development will become important subjects in macroeconomic development. <u>As the first industry of the emerging industries, the development of water and environmental protection enterprises will conform to the current direction of the country's macroeconomic development, and move toward the development of a better investment</u>, while enterprises should also establish effective reform ideas, so as to adapt to the rapidly changing environment and grasp new opportunities brought about by the economic reforms.

改译：It is expected that in the coming years, improving innovation capability, promoting the transformation of economic development mode, and protecting the environment to achieve sustainable development will become important subjects in macroeconomic development. <u>Water and environmental protection enterprises, as the first industry of all emerging industries, with their industrial development targets conforming to the current macroeconomic development direction of China, will meet a better investment and development period.</u> Meanwhile, enterprises should also establish effective reform ideas, to adapt to the rapidly changing environment and grasp new opportunities brought about by the economic reforms.

分析：本段总共两句，都是长句。其中第一句属于长主语句，难度不大。第二句加下画线部分，则是个复杂句，不仅原文难理解，英译时也容易出错。这部分的意思相当于：

作为新兴产业之首的水务环保企业，其行业发展目标顺应了目前中国的宏观经济发展方向，将迎来更好的投资发展期。

原译文跟着原文走，用 as the first industry of the emerging industries 对"水务环保企业"做出说明，但后面的主语用得不对。改译后的主语才是对的。

例 5.63：

原文：在发动机运行时检测各种信号参数和波形：学员可利用万用表、示波器等设备来检测台架电路，学习汽车电路检测。<u>用万用表测量台架面板上电脑端子、发动机电路终端元器件如各种传感器、执行器的电压，电流，阻值，频率等信号参数</u>；用示波器测量各种传感器、执行器的电路波形以及点火部分的高压波形等，学习发动机电路的检测。对照相关的维修手册，可得出这些参数和波形是否合标准。

译文：Check various signal parameters and wave forms while running the engine: the trainees may utilize such equipment as multimeter and oscilloscope to examine chassis circuits and learn how to check car circuits; <u>use a multimeter to measure the voltages, currents, resistances, frequencies and other signal parameters of computer terminals on chassis panels and such engine circuit terminal components as various sensors and actuators;</u> and oscilloscope may be used to measure circuit waveforms of various sensors and actuators and high voltage waveforms of ignition parts and help trainees to learn to check the circuits of the engine. Based on relevant maintenance manual, it is allowed to verify whether these parameters and waveforms meet the standard.

分析：注意，原文加下画线部分的主干是"（学员可）用万用表测量……等信号参数"。

例 5.64：

原文：太阳能电池组件金属边框通过 1×4mm2 铜绞线与支架做螺栓连接，利用边框作为防雷接闪器，金属支架作为接地引下线的方式进行防雷保护，光伏组件场区内，不再设置独立的避雷针保护装置。

原译：Bolted connection is achieved by $1 \times 4\text{mm}^2$ copper stranded wires and brackets of the metal frame of solar cell module, using frame as lightning receptor, metal brackets as grounding downleads to prevent lightning. No separate protective devices of lightning rod will be set inside the PV module area.

改译：The metal frame of solar modules are bolted to the brackets by $1 \times 4\text{mm}^2$ copper stranded wires, with the frame as lightning receptor, metal brackets as grounding downleads to prevent lightning. No separate lightning arrester will be set inside the PV module area.

分析：本例是个复杂句，原译的问题点有：

（1）最主要问题点，是没有理解加下画线的真正意思，其真正意思是"金属边框与支架用螺栓连接起来"。

（2）其次，英文中 bolted connection is achieved 这样的表达，就相当于把中文的"做螺栓连接"说成"螺栓连接被实现了"，其实"做螺栓连接"就是"栓接"的意思。

所以，正确的译法，是在理解清楚相互连接的两个对象（"金属边框"和"支架"）后，再找出句子中真正可以在英文中做谓语动词的中文动词"做螺栓连接"（即"栓接"），并译成 bolted（作动词用）。

例 5.65：

原文：自体免疫细胞治疗技术主要用于癌症治疗和病毒病治疗，是指采集患者自身免疫细胞，<u>经过体外培养、增殖，再回输到病灶</u>，<u>达到杀灭癌细胞和清除病毒目的</u>的治疗技术。

译文：Autologous immune-cell therapy is mainly used in the treatment of cancers and virus diseases, it is a therapy technology to kill cancer cells and remove viruses by gathering autologous immune cells of the patient for invitro culture and proliferation and feeding back to the nidus.

分析：本例实际上相当于两个句子，第一句是原文第一小句，第二句是除第一小句以外的部分。本例翻译的难点在于第二小句。对第二小句的翻译分析如下：

（1）句子主语与第一小句相同，也是"自体免疫细胞治疗技术"，但可用 it 代替。

（2）句子宾语是"治疗技术"，目的状语是"达到杀灭癌细胞和清除病毒目的"，因此译成 a therapy technology to kill ...

（3）"采集患者自身免疫细胞，经过体外培养、增殖，再回输到病灶"这一部分是句子的状语，说明这种治疗技术的实现方法，因此可以译成 by gathering ... for ... and feeding back to the nidus。

本例的翻译，实际上也要先确定句子主干以及其他成分在句子中所扮演的角色，然后先译主干，接着把其他成分的译文放到合适位置，最后再检查、修改和调整句子。

例 5.66：

原文：（该部门）<u>聚焦于攻克恶性、难治性及复发肿瘤的迫切需要</u>，<u>以开展成熟的免疫细胞生物治疗为突破口</u>，建立了免疫生物治疗的适应证，优化治疗方案，有效开展临床患者的随机双盲生物治疗，提高患者的<u>生存质量和延长生存期</u>，并对其中的免疫效应和调控机制开展研究。

译文：<u>Due to</u> the urgent need of treatments for malign, refractory tumors and recurrent tumors, the Department focuses on establishing indications of immune cell-based biotherapies, optimizing therapeutic regimens, conducting effective random double-blind bio-therapies in clinical practice, increasing life quality and lifetimes of patients and studying related immunological effects and regulatory mechanisms <u>based on</u> execution of mature immune cell-based biotherapies.

分析：本例原文句式结构较为复杂，译文总体翻译得比较到位。该译文的亮点如下：

（1）本句表面上看是将"聚焦于"理解成"由于"，译成 due to ... 但实际上，原文的"聚焦于"的宾语不是"攻克……的迫切需要"，而是后面的"建立了免疫生物治疗的适应证，优化治疗方案，有效开展……提高……并对其中的免疫效应和调控机制开展研究。"所以，译文将这些动作都译成动名词形式，并在前面加上 focus on。

（2）将"以……为突破口"理解成"基于……"，译成 based on ...

（3）将"免疫细胞生物治疗"理解成"基于免疫细胞的生物治疗"，译成 immune cell-based biotherapies。

英译中长句

例 5.67：（分类情况的说明）

原文：The durability of prestressed members may be more critically affected by cracking. In the absence of more detailed requirements, <u>it may be assumed that</u> limiting the calculated crack widths to the values of Wmax given in Table 7.1N, under the frequent combination of loads, <u>will generally be satisfactory</u> for prestressed concrete members.

译文：预应力构件的耐用性可受到裂纹的更严重影响。在缺乏更详细要求时，<u>可这样假设，在频遇性荷载组合情况下，将裂纹宽度计算值限制为表 7.1N 中给定的值 Wmax，这样的假设一般可满足预应力混凝土构件的要求</u>。

分析：本例的难点在于第二句。第二句是 it 作形式主语的句子，其主干是加下画线部分，其中 given in Table 7.1N 是 Wmax 的定语，而 under the frequent combination of loads 则是插入语，是 limiting 的条件状语。

例 5.68：

原文：During the course of installation, the Engineer shall have full right for making tests and inspection of the work, as he may deem necessary always with the participation of the Employer's personnel in all tests at Site if so requested by the Employer for the purpose of on-the-job training.

译文：在安装过程中，当工程师认为必要时，工程师具有对工程进行试验与检验的充分权利；若发包商提出要求，则发包商可为在职培训之目的让发包商人员全程参与所有现场试验。

分析：本句的难点在后半部分，即 as he may deem necessary always with the participation of ... for the purpose of on-the-job training。其中 as he may deem necessary 里的 he 是指 the Engineer，意思是"当工程师认为必要时"。而 with the participation of ... for the purpose of on-the-job training 则是指发包商可要求发包商人员全程参与现场试验，以监督整个过程。如果理解了这个背景知识，整个句子就不会觉得难了。

例 5.69：

原文：In preparation for acquiring equipment, funding must be approved prior to issuing purchase orders. <u>Prior to starting activity to acquire new equipment a significant effort should be made to locate equipment available in other Nexteer Automotive facilities that may be appropriate for the intended purpose. If appropriate equipment is available, it should be used as an alternative to purchasing new equipment.</u>

原译：在准备设备采购期间，必须获得资金批准方可发出订购单。<u>开始采购新设备前，需花费大量精力确定其他耐世特汽车工厂内适用于预期目的的设备。如有适当设备，应将其作为采购新设备的替代。</u>

改译：在准备采购设备期间，必须获得资金批准方可发出订购单。<u>在开始采购新设备前，必须尽最大努力确定其他耐世特汽车工厂内是否有适用于预期目的的设备。如有适当设备，应将该设备作为替代，而非采购新设备。</u>

分析：本例原译如下画线部分（第二句和第三句）意思表达不大明确。其分析如下：

（1）第二句：

a significant effort should be made to locate equipment 是主句

prior to starting activity to acquire new equipment 是主句的时间状语

available in other Nexteer Automotive facilities 和 that may be appropriate for the intended purpose 都是 new equipment 的定语。

（2）第三句：

an alternative to purchasing new equipment：替代采购新产品的选择。

例 5.70：

原文：The external pressure coefficients C_{pe} for buildings and parts of buildings depend on the size of the loaded area A, which is the area of the structure, that produces the wind action in the section to be calculated.

译文：建筑或建筑部件的外部压力系数 C_{pe} 与荷载面积 A 的大小相关，荷载面积 A 为对待计算截面产生风力作用的结构物的面积。

分析：本例实际上只有一个句子，其中：

（1）The loaded area A 后面的两小段，即 which is the area of the structure 和 that produces the wind action in the section to be calculated 都是 the loaded area A 的定语从句，其中前一个定语从句 which is the area of the structure 以插入语的形式插入在后一个从句前面。这是英文句子里非常常见的一种句子形式。

（2）and 连接的前后两词，性质完全相同，根据上下文语境，不宜把 and 译成"和""及"，而是要译成"或"。

（3）depend on 用于表示某个参数与另一个或多个参数之间的关系时，通常不要译成"取决于"，而是译成"与……相关"，这是一个数学术语。

例 5.71：

原文：We also hope that we will succeed in demonstrating the astounding versatility of these methods **as reflected in** a number of international case histories from recent years. The reader will hopefully also realise that vibro compaction and vibro replacement stone columns are presently fulfilling – and will continue to do so – all the requirements of sustainable construction. We will demonstrate in a separate chapter of the book that these methods require only natural materials for their execution, leaving behind only a minimal carbon footprint.

原译：希望我们能根据最近几年的国际案例记录，清楚地向读者展现出这些方法的惊人特点。读者也希望能接触到有关于振冲压实和振冲更换碎石桩的令人满意的描述,并且对可持续建筑的要求等方面的事项也抱有同样希望。我们将在书中单独留出一章的篇幅,用以说明,要实施这些方法只需要天然的原材料,而且碳排放量很低。

改译:我们还希望能像最近几年的若干国际案例记录所展示的一样,成功地向读者展现出这些方法惊人的多功能性。<u>但愿读者能认识到,振冲压实与振冲转换碎石桩正在满足,而且将继续满足持续施工的所有要求。</u>我们将在书中单独留出一章的篇幅,用以说明,要实施这些方法只需要天然的原材料,而且可让碳排放达到最少。

分析:本例三个句子的结构分别如下:

(1) We also hope that we will succeed in demonstrating the astounding versatility of these methods **as reflected in** a number of international case histories from recent years.

在句式结构方面,本句的主句很短,是 We also hope that。但从句较长,从句的主句是 will succeed in demonstrating the astounding versatility of these methods,后面的 as reflected in ... 是"如……所体现的一样",as reflected in 后是有较长修饰的名词。在用词方面,as reflected 的本意是"如……所展示",但根据语境,宜将 as reflected 译成"像……展示的一样",将 histories 译成"记录"。

(2) The reader will hopefully also realise that vibro compaction and vibro replacement stone columns are presently fulfilling – and will continue to do so – all the requirements of sustainable construction.

在句式结构方面,本句主句是 The reader will hopefully also realise that,其中 hopefully 是"但愿"的意思。从句的难点是破折号之后的内容,其中 fulfilling 的宾语是 all the requirements,而 and will continue to do so = and will continue to fulfill all the requirements。也就是说,do so = fulfill。

(3) We will demonstrate in a separate chapter of the book that these methods require only natural materials for their execution, leaving behind only a minimal carbon footprint.

本句的第一小句中，demonstrate 的宾语是 that 引导的宾语从句，in a separate chapter of the book 是地点状语，后面的 leaving behind only a minimal carbon footprint 则是现在分词短语做状语，起到补充说明的作用。

例 5.72：（表达方式的英文 in such a way）

原文：All work shall be carried out in such a way as to allow access and afford all reasonable facilities for any other Contractor and his workmen and for the workmen of the Employer and any other person who may be employed in the execution and/or operation at or near the site of any work in connection with the Contractor or otherwise.

原译：应以此方式进行施工，以便其他承包商及其施工人员，业主的施工人员以及被雇用于施工和/或操作的处于或靠近工地的其他相关人员能使用及负担得起适用设施。

改译：工程开展方式应使得其他承包商及其工作人员、业主工作人员，以及在施工和/或作业过程中可能雇佣的、处于或靠近与本承包商相关的任何工程现场的任何其他人员能使用及负担得起适用设施。

分析：本句是一个典型的英文长句，其句子结构如下：

All work shall be carried out in such a way as to allow access and afford all reasonable facilities for any other Contractor and his workmen and for the workmen of the Employer and any other person who may be employed in the execution and/or operation at or near the site of any work in connection with the Contractor or otherwise.

其中的句子层次如下：

（1）第一层次是 All work shall be carried out in such a way as to allow ... 是全句的主句。

（2）第二层次是 allow access and afford all reasonable facilities for ... and for ... 这一层次后面的两个 for 后的名词是两个动词 access 和 afford 的施动者。

（3）第三层次是 any other person 后面的定语从句 who may be employed。

（4）第四个层次以后的内容是 in the execution and/or operation at or near the site of any work in connection with the Contractor or otherwise，是 employed 的时间状语、地点状语以及方式状语。其中 or otherwise 是与 in the execution and/or operation 相并列的成分，意思是"或以其他方式／其他情况下（被雇佣）"，在这里可以省译。

2. 分点表达的句子

例 5.73：

原文：The EPC Contractor shall furnish all designs, materials, equipment and supply, all labour necessary for the complete installation, testing and operation of the plumbing and sanitary systems required in the Employer's Camp and Offices, Switchyard and in the Powerhouse, including but not limited to the following:

(a) Potable water systems including treatment plant.

(b) Water service connections.

(c) Plumbing and sanitary fixtures, fittings, trims and accessories.

(d) Sewage, waste and vent pipe systems.

...

译文：统包总承包商应为雇主宿舍和办公室、开关站和发电站要求的管道和卫生系统的完整安装、测试和操作提供所需的设计、材料、设备、日用品以及人力，这些系统包括但不限于：

（a）饮用水系统，包括水处理厂。

（b）供水装置接入系统。

（c）管道与卫生设备、配件、装饰及附件。

（d）污水管、废水管与通气管系统。

……

分析：本例由于主句后面有好几个分点说明的内容，所以，最后面的 including but not limited to the following 不宜简单地译成"包括但不限于"，而是要译成"这些系统包括但不限于"，否则就可以理解成"设计、材料、设备、日用品以及人力"包括但不限于以下几点。

例 5.74：

原文：Maximum continuous current carrying capacity (Amps)

(a) When laid direct in ground at 140 cm depth from kerbstone or final ground level to the lowest point of the cable surface and the cables spaced 12cms, centre to centre ground temp. 35℃ g=120 for one 3 phase circuit per trench.

(b) When drawing into pipes or ducts of length more than 15 metre for one 3 phase circuit.

原译：最大连续载流量

（a）电缆直埋在地下 140 cm 深或当地平面距离电缆表面最低点和电缆的距离为 12 cm，中心到中心地面温度为 35℃，g=120 时每个电缆沟中的一条 3 相电路的最大连续载流量。

（b）装入长度超过 15 米的管道或电缆槽导管时的一条 3 相电路。

改译：最大连续载流量

（a）当以路缘石或最终地平面到电缆表面最低点的距离计算深度，电缆的铺设深度为 140cm，且电缆中心距为 12cm，地面温度为 35℃，g=120 时，每个电缆沟中的一条 3 相电路的最大连续载流量。

（b）装入长度超过 15 米的管道或导管时，一条 3 相电路的最大连续载流量。

分析：本句的主干实际上是：

Maximum continuous current carrying capacity (Amps) (of cables)

(a) when ... for one 3 phase circuit per trench;

(b) when ... for one 3 phase circuit.

而且，主干相当于省略了 of cables，两个条件状语从句也有所省略，句子的完整表达应为（加下画线部分为省略成分）：

Maximum continuous current carrying capacity (Amps) <u>of cables</u>

(a) When <u>cables are</u> laid direct in ground at 140 cm depth from kerbstone or final ground level to the lowest point of the cable surface and the cables <u>are</u> spaced 12cms, centre to centre ground temp. <u>are</u> 35℃ <u>,</u> g=120 for one 3 phase circuit per trench.

(b) When drawing the cables into pipes or ducts of length more than 15 metre for one 3 phase circuit.

也就是说，（a）（b）两小点其实是对前面小标题（主干）的分点说明。因此，在组织译文时，宜将（a）（b）两点的译文都跟前面的小标题（主干）相呼应，译成"……时……的最大连续载流量"。

例 5.75：

原文：The services to be re-performed shall commence immediately following the loss or damage in question and shall end either:

When the hole, well or casing is restored to the condition it was in immediately prior to the loss or damage; or

When a replacement well reaches the same depth and is in the same condition as the old well immediately prior to the loss or damage; or

Where restoration or re-drilling is impossible, when the well in question has been properly abandoned in accordance with good oilfield practice.

译文：如存在问题的井已经根据良好油田惯例予以妥善废弃，且不可能恢复或重新钻井，则服务的重新实施应在所述损失与损害之后立即进行，并在达到下列任一状态时予以终止：

井眼、井身或套管恢复至紧接着该损失或损害发生之前的状态；或

备用井达到旧井在紧接着损失或损害之前的相同深度与相同状态；或者

存在问题的井已经根据良好油田惯例予以妥善废弃，且不可能恢复或重新钻井。

分析：在本例中，有些人可能会对第一小段主句中的 either 的使用存在疑问，因为 either 通常只使用在两个选项的情况下，但这里却有三个选项。但是，细心的译员可能会发现，最后一小段前面虽然加点，但句式明显与前面第二和第三小段不同。再仔细想想，就会发现，其实最后一小段前面不该加点，应该与第一小段同样都是主句。经过这样理解后，译员就应该知道怎么处理这四个小段了（应把最后一小段译文作为条件状语从句提到第一小段，而最后一小段译文则留空）。

3. 术语与定义

例 5.76：

原文：① 防碰天车制动

在钻机防碰天车系统给出信号后，盘刹系统自动实施的安全保护制动。通常该制动是由工作钳和安全钳共同执行。

原译：Crown-block protecting brake

Crown-block protecting brake, generally actuated jointly by the service caliper and safety caliper, is automatically conducted when the crown-block protector system of rig sends signal.

改译：Crown-block protecting brake

Safety protecting brake automatically actuated by disk brake system after the crown-block protector system of the rig gives signals, generally jointly actuated by the service caliper and safety caliper.

分析：本例是文件中对术语"防碰天车制动"的解释。像这种解释，通常要把译文组织成非句子形式，即以解释说明中的核心词作为英文的核心词，其他内容全部作为其修饰成分。本例的定义看起来似乎是两个句子，其实第一个句号之前不是句子，其本质还是个名词，而后面一小句在语法上算是个独立句子。但因为原文是对术语的定义，所以，还是应该将第一个句号改成逗号，并将整个定语改成"……的安全保护制动，通常由工作钳和安全钳共同执行"。这样修改后，就可以将译文译成参考译文那样的名词形式（非句子形式）了。

例 5.77：

原文：Anaerobic digestion

Biological degradation of organics (eg., food waste and green garden waste) in the absence of oxygen, producing biogas suitable for energy generation (including transport fuel), and residue (digestate) suitable for use as a soil improver.

Carbon dioxide (CO_2)

Carbon dioxide is a naturally occurring gas comprising 0.04 percent of the

---

① 本例原文选自中国石油天然气行业标准 SY/T6727-2008。

atmosphere. The burning of fossil fuels releases carbon dioxide fixed by plants many millions of years ago, and this has increased its concentration in the atmosphere by some 12 percent over the past century. It contributes about 60 percent of the potential global warming effect of man-made emissions of greenhouse gases.

原译：厌氧消化

食品垃圾和园林绿地废物等有机物，在缺氧的情况下会降解产生能源和残渣（包括运输燃料），残渣可以当作肥料改善土壤。

二氧化碳（CO_2）

二氧化碳是一种自然物质，占大气成分的 0.04%。数百万年前燃烧化石燃料产生的二氧化碳通过植物吸收固化，过去 100 年里，大气中的二氧化碳含量浓度增加了 12%。在影响全球变暖的气体中，人为排放的二氧化碳占 60%。

改译：厌氧消化

有机物（如食品垃圾与绿色园林废物等）在缺少氧气的条件下的生物降解，此类降解可生成能产生能量（包括交通燃料）的沼气，且其残渣（沼渣）可以当作土壤改善剂。

二氧化碳（CO_2）

二氧化碳是一种自然物质，占大气成分的 0.04%。在数百万年前被植物固定的二氧化碳，在过去 100 年里因燃烧化石燃料被释放出来，这使大气中的二氧化碳含量浓度增加了 12%。在影响全球变暖的气体中，二氧化碳占 60%。

分析：注意，在标准规范类或其他说明类文章中，通常文章的前面会有一些术语解释。这些术语解释，通常是把术语单独列为一行，后面紧跟一小段对术语进行解释说明。通常情况下，不管是中文还是英文，这种解释说明性段落都不是一个完整的句子，而只是相当于一个名词。因此，在翻译时，应该译成"……的XX"（直接详细说明特征，并在最后说明其属性），或者"……的XX……"（先简要说明概念或属性，然后再补充说明）。

例 5.78：（借助网络定义，对译文进行修改）

原文：The International Organization for Standardization's (ISO) 14001 environmental management system standard defines an organization as a company, corporation, firm, enterprise, authority, or institution, or part or combination thereof, whether incorporated or not, public or private, that has its own functions and administration.

原译：国际标准化组织（ISO）14001 环境管理体系标准将组织定义为有自身职能与行政管理的公有或民营的公司、团体、商号、企业、管理机构、事业单位或者以上各团体的其中之一部分或组合（包括法人团体及非法人团体）。

改译：国际标准化组织（ISO）《14001 环境管理体系标准》将组织定义为具有自身职能与行政管理的公司、集团公司、商行、企业、政府机构或事业单位，或者以上各单位的一部分或结合体，无论其是否法人团体、公营或私营。

分析：这一段的意思理解基本没问题，但表达方面，总感觉不是很好。想借鉴 ISO 对"组织"的定义，但一直不容易查到。在百度"ISO 组织定义""ISO 组织是什么"都查不到想要的，而查"ISO14001 术语 组织 定义"，结果如下：

其中第一条结果前面部分如下：

为便于理解和把握，我们把ISO14001标准中的13个定义按相互关系分为五组。

一、环境---环境因素---环境影响

环境 environment：
组织运行活动的外部存在，包括空气、水、土地、自然资源、植物、动物、人，以及它们之间的相互关系。
注：从这一意义上，外部存在从组织内延伸到全球系统。

环境因素 environmental aspect：
一个组织的活动、产品或服务中能与环境发生相互作用的要素。
注：重要环境因素是指具有或能够产生重大环境影响的环境因素。

环境影响 environmental impact：
全部或部分地由组织的活动、产品或服务给环境造成的任何有害或有益的变化。

二、组织---相关方

组织 organization：
具有自身职能和行政管理的公司、集团公司、商行、企事业单位、政府机构或社团，或是上述单位的部分或结合体，无论其是否法人团体、公营或私营。
注：对于拥有一个以上运行单位的组织，可以把一个运行单位视为一个组织。

相关方 interested party：
关注组织的环境表现（行为）或受其环境表现（行为）影响的个人或团体。

这里对于"组织"的定义，句子更为顺畅，因此对原译文进行如上相应修改。

六、逻辑及背景知识

在应用类翻译中，有些句子的原文语法可能不大合理，但译员对句子所涉背景知识很熟悉时，通常可以从逻辑上对句子加以理解。但是，如果译员不熟悉相关专业背景知识，一时难以透彻理解时，则需要通过上下文隐含的背景知识加以理解，或者利用上下文的相关专业词汇，到网络或专业书籍上了解相关背景知识，从而透彻理解原文。本书前文已经有些句子运用到逻辑和背景知识进行翻译，这里将再举些例子，以便大家加深印象。

（一）利用逻辑理解

例 5.79：

原文：Electrocution occurs when the animal connects two phases or a phase to earth.

原译：当动物同时接触两种电位或同时接触一种电位和地表时，会发生触电。

改译：当动物同时接触两条相线或同时接触相线和大地时，就会触电。

分析：本句里的 connect，本义是"连接"，但根据语境和逻辑推理，宜将其译为"接触"，而且，connect 的宾语有两部分，第一部分是 two phases，第二部分是 a phase to earth，分别构成 connect two phases 和 connect a phase to earth 这样的表达。

例 5.80：
原文：

PN25 Flanges

Nominal Diameter	Flange O.D. (mm)	Bolt Circle Dia. (mm)	Bolt Hole Dia. (mm)	Bolt Number	Bolt Dia.
200	260	310	28	12	M24
300	485	432	31	16	M27
400	620	550	36	16	M33
600	845	770	39	20	M36
800	1085	990	49	24	M45
1000	1320	1210	56	32	M52
1200	1530	1420	56	32	M52

原译：

PN25法兰

公称直径	法兰外径（mm）	螺栓圆直径（mm）	螺栓孔直径（mm）	螺栓编号	螺栓直径
200	260	310	28	12	M24
300	485	432	31	16	M27
400	620	550	36	16	M33
600	845	770	39	20	M36
800	1085	990	49	24	M45
1000	1320	1210	56	32	M52
1200	1530	1420	56	32	M52

改译：

PN25法兰

公称直径	法兰外径（mm）	螺栓圆直径（mm）	螺栓孔直径（mm）	螺栓个数	螺栓直径
200	260	310	28	12	M24
300	485	432	31	16	M27
400	620	550	36	16	M33
600	845	770	39	20	M36
800	1085	990	49	24	M45
1000	1320	1210	56	32	M52
1200	1530	1420	56	32	M52

分析：Bolt number 是"螺栓个数"而不是"螺栓编号"。因为这里的数字都是 4 或 8 的倍数，是不连续的。而且，法兰的螺栓一般都是两两相对的，一般都是 4 或 8 的倍数。如下图所示：

两个法兰盘扣到一起，并在洞里装上螺栓旋紧，就可以把两节管道给连接起来。如下图：

例 5.81：

原文：现场电气线路，必须按规定架空敷设坚韧橡皮线或塑料护套线。<u>在通道或马路处可采用加保护管埋设地下</u>，树立标志，<u>接头必须架空设接头箱</u>。

译文：Onsite electrical circuit must be laid overhead tough rubber wires or plastic-sheathed wires as specified. <u>The wires crossing a passageway or a road may be buried underground in protective tubes</u>, with signs set up; <u>the joints must be installed in an overhead box</u>.

分析："加保护管埋设地下"的意思，是指将电线装在管道中埋在地下，所以译成 be buried underground in protective tubes。"架空设接头箱"是指接头必须架空，并在完成接线后置于接头箱中。

例 5.82：

原文：The recoverable reserves are approximately 52x109 Sm3 and the maximum production rate is 20.7 MSm3/d. The rich gas is exported to shore through a 150 km long 30" pipeline with a design capacity of 26.5 MSm3/d, which is parallel to the Troll pipeline <u>the last 65 km</u> towards shore.

原译：该气田可采储量大约为 52x109 Sm3，最大产量为 20.7MSm3/d。富气通过 150km 长的 30" 管线输送至陆地，管线设计能力为 26.5MSm3/d，管线与 <u>65km</u> 长的 Troll 管线平行，通往陆地。

改译：该气田可采储量大约为 52×10^9 Sm^3，最大产量为 20.7MSm^3/d。富气通过 150km 长的 30" 管线输送至陆地，管线设计能力为 26.5MSm^3/d，<u>其最后 65km</u> 长的管线与 Troll 管线平行，通往陆地。

分析：本句难点在最后一小句的 the last 65 km。译员不理解原文想表达的意思，认为 last 是多余的。但通过逻辑可以知道，这一小句的意思是该管线最后 65km 与 Troll 管线平行，最后通往陆地。另外，Sm3 其实是 Sm^3，指的是标准状态（我国指导 20℃、标准大气压状态）下的气体体积（立方米）。

例 5.83：

原文：The replacement of exhausted oils must be carried out <u>with warm machine</u>, immediately after a stop.

原译：废油更换必须<u>在运行的机器中进行</u>，在机器停止后立即进行。

改译：废油更换必须在机器停止后，<u>还处于温热状态时</u>立即进行。

分析：原译文逻辑明显有问题。warm machine 是指刚停机时处于温热状态的机器（热机状态）。之所以要在热机状态下更换废油，是因为油在温热状态更软，甚至处于液态，易于除掉旧油、加上新油。

例 5.84：

原文：When a BOP, component, or assembly is taken out of service for an extended period of time, it shall be completely washed or steam cleaned, <u>and machined surfaces coated with a corrosion inhibitor.</u>

For BOPs, the rams or sealing element shall be removed and the internal BOP body/cavities shall be thoroughly washed, inspected, and coated with a corrosion inhibitor in accordance with the equipment owner's and manufacturer's requirements.

原译：如防喷器、部件或组件被取出并长期不再使用，应彻底清洗或用蒸汽清洗，<u>并对被缓蚀剂包覆的表面进行清理</u>。

对于防喷器，其闸板或密封元件应根据设备所有者或制造商的要求进行拆卸，且内部防喷器体/腔应进行彻底清洗、检验，并涂上缓蚀剂。

改译：如防喷器、部件或组件被取出并长期不再使用，应彻底清洗或用蒸汽清洗，<u>并将加工面涂上缓蚀剂</u>。

对于防喷器，其闸板或密封元件应根据设备所有者或制造商的要求进行拆卸，且内部防喷器体/腔应进行彻底清洗、检验，并涂上缓蚀剂。

分析：第一段最后一小句让人感觉不好理解。根据逻辑推理，可以认定原译"并对被缓蚀剂包覆的表面进行清理"理解不对，应该改成"并将加工面涂上缓蚀剂"。

（二）利用背景知识理解

例 5.85：

原文：Gamma ray logs are effective in distinguishing permeable zones because radioactive elements tend to be concentrated in the shales, which are impermeable, and are much less abundant in carbonates and sands, which are generally permeable.

原译：伽马射线测井有助于区分渗透性带，这是因为放射性元素在页岩中会聚集在一起，而页岩是不渗透的，其含有的碳酸盐和砂（一般具有渗透性）较少。

改译：伽马射线测井有助于区分渗透性带，这是因为放射性元素会聚集在页岩（不具有渗透性）中，而在碳酸盐岩和砂岩（一般具有渗透性）中的含量较低。

分析：本句原译文最后一小句不大顺畅（"其含有的碳酸盐和砂[一般具有渗透性]较少"逻辑上说不通）。因此，合理怀疑其理解可能存在问题。从句子结构上看，原文的意思可能是"放射性元素会聚集在页岩中，但在 carbonates 和 sands 中含量较低"。但是 carbonate 和 sand 分别是"碳酸盐"和"砂"的意思，这时，我们可推测 carbonate 和 sand 分别是"碳酸盐岩"和"砂岩"。这样，本句就可以理解成"放射性元素会聚集在页岩中，但在碳酸盐岩和砂岩中含量较低"。为证实这一逻辑的准确性，我们还必须到网络上查看一下有没这样的知识点（即"放射性元素在碳酸盐岩和砂岩中含量较低"）。在百度里查"碳酸盐岩 放射性元素"，结果如下：

从搜索结果中，发现好几条链接里都有"碳酸盐岩地层中自然放射性元素含量低"的说明。在网络上还可证实砂岩中自然放射性元素含量较低，但相对没那么容易查到。而如果查看《矿物地球物理》（石油大学出版社）第七章"自然伽马测井和放射性同位素测井"第二节"自然伽马测井"中的"岩石的自然放射性"，则可以证实，砂岩中的自然放射性元素的含量确实较低。

综上所述，本例的改译译文才是对原文的准确理解。

例 5.86：

原文：In many models (e.g. Kyllingstad & Halsey, 1988 and Brett, 1992) the natural oscillation of the drill string is represented very simply – for example via a single degree of freedom system in which one third of the polar moment of inertia of the compliant part of the drill string (Jc) is added to the inertia of the essentially rigid part (Jr) so that the undamped natural frequency

$$f_0 = \frac{1}{2\pi}\sqrt{\frac{K_t}{J}},$$

where Kt is the overall torsional stiffness and $J = J_r + J_c/3$,

In the real system there will be many modes of oscillation and corresponding natural frequencies.

译文：在许多模型（如：Kyllingstad 和 Halsey，1988，以及 Brett，1992）中，钻柱的自然振荡表现得非常简单——例如，通过单自由度体系，将钻柱的柔性部分（Jc）的三分之一惯性极矩与基本刚性部分（Jr）的惯量相加，以使得无阻尼自然频率

$$f_0 = \frac{1}{2\pi}\sqrt{\frac{K_t}{J}}$$

其中：kt 为总抗扭刚度，$J = J_r + J_c/3$，

在实际体系中，会有许多种振荡模式和相应的自然频率。

分析：本例涉及数学、采油和物理等专业知识。其中的相关专业词汇有 model（模型）、oscillation（振荡）、drill string（钻柱）、degree of freedom（自由度）、polar moment of inertia（惯性极矩）、compliant part（柔性部件）/ essentially rigid part（基本刚性部件）、inertia（惯量）、undamped natural frequency（无阻尼自然频率）、torsional stiffness（抗扭刚度）等。要译好这段话，必须具有相关专业知识或者相关翻译经验。

例 5.87：

原文：Heavily weathered coral is defined as a coral layer with SPT N-values being greater than or equal to 50 **blow**.

译文：强风化珊瑚被定义为标准贯入试验 N 值大于或等于 50 击的珊瑚层。

分析：本句的难点是 blow 这个词的译法。

本句中的 SPT 可以查到是"标准贯入试验"，SPT N 是"标准贯入试验 N 值"。因此，可以在百度里查"标准贯入试验"，点击第一条"标准贯入试验＿百度百科"，其中有以下内容：

操作

标准贯入试验多与钻探相配合使用，操作要点是：

①钻具钻至试验土层标高以上约15厘米处，以避免下层土受扰动。

②贯入前，应检查触探杆的接头，不得松脱。贯入时，穿心锤落距为76厘米，使其自由下落，将贯入器直打入土层中15厘米。以后每打入土层30厘米的锤击数，即为实测(锤击数N)。

③提出贯入器，取出贯入器中的土样进行鉴别描述。

④若需继续进行下一深度的贯入试验时，即重复上述操作步骤进行试验。

⑤当钻杆长度大于3米时，(锤击数)应按下式进行钻杆长度修正：N63.5=αN，式中N63.5为标准贯入试验锤击数，α为触探杆长度校正系数，如触探杆长分别为≤3、≤6、≤9、≤12、≤15、≤18、≤21米时，则α相应分别为1、0.92、0.86、0.81、0.77、0.73、0.70。

标准贯入试验

测定

根据标准贯入试验锤击数测定各类砂的地基承载力（公斤/平方厘米），一般为：

①当(击数)大于30时，密实的砾砂、粗砂、中砂（孔隙比均小于0.60）为4公斤/平方厘米；

②当(击数)小于或等于30而大于15时，中密的砾砂、粗砂、中砂（孔隙比均大于0.60而小于0.85）为3公斤/平方厘米，细砂、粉砂（孔隙比均大于0.70而小于0.85）为1.5—2公斤/平方厘米；

③当(击数)小于或等于15而大于或等于10时，稍密的砾砂、粗砂、中砂（孔隙比均大于0.75而小于0.85）为2公斤/平方厘米，细砂、粉砂（孔隙比均大于0.85而小于0.95）为1—1.5公斤/平方厘米。对于老粘土和一般粘性土的容许承载力，当(击数)分别为3、5、7、9、11、13、15、17、19、21、23时，则其相应的容许承载力分别为1.2、1.6、2.0、2.4、2.8、3.2、3.6、4.2、5.0、5.8、6.6公斤/平方厘米。[1]

因此，确定 blow 为"击数"或"击"。

例 5.88：

原文：Ferrous/Non-Ferrous Metal Production

原译：含铁金属 / 无铁金属生产

改译：黑色金属 / 有色金属生产

分析：ferrous metal：黑色金属；non-ferrous metal 有色金属。

查找百度百科的词条"有色金属"，可以得知关于黑色金属和有色金属的定义。

黑色金属和有色金属这名字，常常使人误会，以为黑色金属一定是黑的。其实，黑色金属只有三种：铁、锰、铬。而它们三个都不是黑色的！纯铁是银白色的；锰是银白色的；铬是灰白色的。因为铁的表面常常生锈，盖着一层黑色的四氧化三铁与棕褐色的三氧化二铁的混合物，看上去就是黑色的。所以人们称之为"黑色金属"。常说的"黑色冶金工业"，主要是指钢铁工业。因为最常见的合金钢是锰钢与铬钢，这样，人们把锰与铬也算成是"黑色金属"了。

除了铁、锰、铬以外，其他的金属，都算是有色金属。

在有色金属中，还有各种各样的分类方法。比如，按照比重来分，铝、镁、锂、钠、钾等的比重小于 4.5，叫做"轻金属"，而铜、锌、镍、汞、锡、铅等的比重大于 4.5，叫做"重金属"。像金、银、铂、锇、铱等比较贵，叫做"贵金属"，镭、铀、钍、钋等具有放射性，叫做"放射性金属"，还有像铌、钽、锆、铷、金、镭、铪、铀等因为地壳中含量较少，或者比较分散，人们又称之为"稀有金属"。

以上是冶金行业的最基本知识，希望大家能记住，如果实在记不住，至少要知道黑色金属和有色金属的英文。

例 5.89：

原文：AC connection type Screw terminal block, cable gland PG36

<u>Three phase 3W or 4W+PE</u>

译文：交流连接类型螺钉端子接线盒、电缆接头 PG36

<u>三相三线或四线 + 保护接地</u>

分析：在电学中，各缩写符号的意思如下：

W = wire

PE = protective earthing

N = neutral

P = phase

本节课后练习二

翻译下列句子，注意下画线部分：

1. 在规定条件下粉碎样品，不同硬度的小麦穿过筛网的粉粒<u>质量</u>是不同的，用留存在筛网上的粉粒<u>质量</u>占测试小麦样品质量的百分比值表征小麦的硬度<u>指数</u>，简称 HI。

2. 多束伽马射线集中照射于预选的靶点，可产生一个伽马射线的高剂量区，从而一次或几次、致死性地摧毁置于<u>焦点</u>的病变组织，以达到外科手术切除或损毁<u>病灶</u>的效果。

3. 半导体工业用石英玻璃作为高温容器或扩散管时，由于半导体材料要求很高的纯度，所以要求与石英玻璃接触的作为炉衬的耐火材料必须预先经过高温和清洁处理，除掉钾、钠等碱性杂质，然后才能放入石英玻璃内使用。

4. 中国科学技术大学郭光灿院士领导的中科院量子信息重点实验室在高维量子信息存储方面取得重要进展，该实验室史保森教授领导的研究小组在 2013 年首次成功地实现了携带轨道角动量、具有空间结构的单光子脉冲的存储与释放，证明了高维量子态的存储是完全可行的。

5. The silicone encapsulation is also vulnerable to nonpolar fluids and solvents commonly used during the manufacturing process of the luminaries such as cleaning, oil assisted drilling, and any processes that would allow the Bridgelux V Series to come into contact with the fluids or solvents.

6. The wire gauge will affect the thermal load placed on the soldering system, so a larger diameter wire (smaller gauge numbers) may require a higher soldering iron temperature setting or a longer soldering cycle time than a smaller diameter wire with the same LED array.

7. Site repairs will not be permitted where the total area of damage to be repaired on a single pipe or fitting exceeds 6cm^2 for pipes and fittings up to and including 600mm diameter and 12cm^2 for pipes and fittings in excess of 600mm diameter.

8. After the series of LiF-AlkF (Alk = Na, K, Rb, Cs) binary alkali salts that have been analysed in recent years the LiF-CaF$_2$ system has been measured highlighting for the first time the heat capacity behaviour in the monovalent-divalent cationic fluoride system.

第四节　回译（返译）

一、讲话的回译

例 5.90：

原文：This report is respectfully submitted in order to update the SOA on the achievements that ConocoPhillips China Inc （"COPC"）and China National Offshore Oil Company （"CNOOC"）, the PL 19-3 co-venturers, have made in regards to the various directives that the SOA have issued since the onset of the oil spill events in June, 2011. Specifically, this report aims to address key concerns raised by the SOA in their communications with COPC in which the SOA states that COPC needs to continue taking further effective measures to implement the requirements of the "Three Continues", specifically to "<u>continue to identify oil spill release sources, isolate and seal the oil release sources, and clean up the released oil in a timely fashion.</u>"

原译：兹将本报告郑重提交给中国国家海洋局[①]，以更新康菲石油中国有限公司（以下称为"康菲中国"）与中国海洋石油总公司（以下称为"中海油"）——即 PL 19-3 的联合投资方——自从 2011 年 6 月溢油事件发生以来由国家海洋局所发布的各项指令有关方面所取得的成就。具体而言，本报告旨在解决国家海洋局与康菲中国的沟通中所提出的关切问题，在这些沟通中，国家海洋局指出，康菲中国必须继续采取进一步有效措施以执行"三个继续"的要求，即及时地"<u>继续及时确定溢油源、及时隔离封堵溢油源、及时清理已泄漏油污</u>"。

改译：兹将本报告郑重提交给中国国家海洋局，以更新康菲石油中国有限公司（以下称为"康菲中国"）与中国海洋石油总公司（以下称为"中海油"）——即 PL 19-3 的联合投资方——自从 2011 年 6 月溢油事件发生以来由国家海洋局所发布的各项指令有关方面所取得的成就。具体而言，本报告旨在解决国家海洋局与康菲中国的沟通中所提出的关切问题，在这些沟通中，国家海洋局指出，康菲中国必须继续采取进一步有效措施以执行"三个继续"的要求，即及时地"<u>继续排查溢油风险点、继续封堵溢油源、继续清理油污</u>"。

分析：本文中 Three Continues 和后面 continue to 的具体内容，一看就知道是正式公告文本。Three Continues 应该是"三个继续"或者"三个连续"什么的，因此，后面的 continue to 的内容，就不应该按自己的想象进行翻译了。

但是，用"三个继续"直接到网络上搜索，很难直接找到跟这个原文相关的内容。

[①] 机构改革前名称。

这说明还必须再缩小搜索范围，由于前面出现了 SOA（国家海洋局），所以，可以搜索"国家海洋局 三个继续"，但好像还不是很容易查到。这时，可以再增加一些相关关键词，搜索"国家海洋局 康菲石油 三个继续"，结果，在百度里，好几个链接在打开后都可以查找到"三个继续"，即"继续排查溢油风险点、继续封堵溢油源、继续清理油污"。

例 5.91：

原文：Huawei has broken out from the pack of smartphone handset makers to become No. 3 in the world, behind only Apple and Samsung, mostly by focusing on low-cost handsets and emerging markets like China, India and Africa.

So the comments from the head of its consumer business group to the *Wall Street Journal* are not going to sit well with Microsoft and Samsung.

Richard Yu said his company is giving up on Windows Phones. "It wasn't profitable for us. We were losing money for two years on those phones. So for now we've decided to put any releases of new Windows phones on hold," he told the Journal.

And Tizen, Samsung's alternative phone OS to Android? "In the past we had a team to do research on Tizen but I canceled it," Yu said. "We feel Tizen has no chance to be successful."

原译：华为的低成本手机策略和专攻中国、印度与非洲等新兴市场，使其从一众智能手机制造商中脱颖而出，一跃成为仅次于苹果和三星的世界第三大智能手机制造商。

正因如此，华为消费者业务集团 CEO 余承东在面对《华尔街日报》采访时给出的说法无法令微软和三星满意。

余承东在《华尔街日报》的采访中透露，华为已经放弃 WP 手机，"WP 手机对华为来说不是一个有利的选择。华为曾在 Windows Phone 手机业务上亏损了两年，现在已经搁置了该机型的发布。"

当被问到三星公司用于替代安卓系统的手机操作系统 Tizen 时，余承东说道，"华为此前曾经有专门的泰泽团队，但我取消掉了，因为我们都认为泰泽不会成功。"

改译：华为的低成本手机策略和专攻中国、印度与非洲等新兴市场，使其从一众智能手机制造商中脱颖而出，一跃成为仅次于苹果和三星的世界第三大智能手机制造商。

正因如此，华为消费者业务集团 CEO 余承东在面对《华尔街日报》采访时给出的说法无法令微软和三星满意。

余承东在《华尔街日报》的采访中透露，华为已经放弃 WP 手机，"它无法给我们带来利润。此类手机曾连续两年给我们造成损失。因此目前我们决定暂时延后所有新款 Windows 系统手机的发布。"

当被问到三星公司用于替代安卓系统的手机操作系统 Tizen 时，余承东说道，"过去我们曾经有一个研究 Tizen 的团队，但我解散了这个团队。"他说，"我们觉得 Tizen 没有成功的机会"。

分析：注意，这几段英文讲的是中国的华为公司高层主管接受采访时所讲的话，这些话，正常情况下肯定是用中文讲的。因为原话是中文，所以在回译时，必须查到原始中文表达，不能自己译。

查找时，可以查"余承东 华尔街日报采访"，结果就可以找到相关原话。如以下链接：http://news.ittime.com.cn/news/news_1799.shtml（2020 年 5 月 30 日检索）。

二、文件名的回译

例 5.92：

原文："PRC GAAP" means the [Basic Standard and 38 Specific Standard of the Accounting Standards for Business Enterprises issued by the Ministry of Finance on 15 February 2006, and the Application Guidance for Accounting Standards for Business Enterprises, Interpretations of Accounting Standards for Business Enterprises] and other relevant regulations issued thereafter;

原译："中国会计准则"指【财政部于 2006 年 2 月 15 日颁布的 1 项企业基本会计准则和 38 项企业具体会计准则，企业会计准则应用指导和企业会计准则解释】及随后颁布的相关规定；

改译:"中国通用会计准则"指【财政部于 2006 年 2 月 15 日颁布的《企业会计准则——基本准则》和 38 项具体准则,以及《企业会计准则——应用指南》和《企业会计准则解释》】,以及随后颁布的其他相关规定;

分析:注意,本例中各项会计相关规定的译法应该采用中文既有表达法,不能自己译,必须到百度或其他文件里查出正确的中文表达法。

 企业会计基本准则　　　　　　　　　　　　　　百度一下

网页　新闻　贴吧　知道　音乐　图片　视频　地图　文库　更多»

百度为您找到相关结果约11,600,000个　　　　　　　　　　▽搜索工具

企业会计基本准则_百度百科

企业会计准则体系包括《企业会计准则——基本准则》(以下简称基本准则)、具体准则和会计准则应用指南和解释等。基本准则是企业会计准则体系的概念基础,是具体准则、应用指南...
地位作用　主要内容
baike.baidu.com/ ▼ - ▷

企业会计准则——基本准则(2014年修正)
2014年7月30日 - 2014年7月23日根据《财政部关于修改<企业会计准则——基本准则>的决定》修改)第一章 总 则 第 一 条 为了规范企业会计确认、计量和报告行为,保证会计信...
www.360doc.com/content... ▼ - 百度快照 - 796条评价

企业会计准则——基本准则_百度百科
2014年4月7日 - 《企业会计准则——基本准则》于2006年2月15日以财政部令第33号公布,根据2014年7月23日中华人民共和国财政部令第76号《财政部关于修改<企业会计准则——...
baike.baidu.com/link?u... ▼ - V₃ - 百度快照 - 评价

中华人民共和国财政部令第76号——财政部关于修改《企业会计准则...
——财政部关于修改《企业会计准则——基本准则》的决定《财政部关于修改<企业会计准则——基本准则>的决定》已经财政部部务会议审议通过,现予公布,自公布之日起...
tfs.mof.gov.cn/zhengwu... ▼ - 百度快照 - 评价

《企业会计准则——基本准则》全文(2014年)_企业会计准则..._会计网
2014年11月29日 - 导读:7月29日,财政部作出对《企业会计准则——基本准则》的修改决定,对《企业会计准则——基本准则》第四十二条第五项进行了修改。...
www.kuaiji.com/fagui/1... ▼ - V₁ - 百度快照 - 617条评价

 企业会计准则应用指南　　　　　　　　　　　　　百度一下

网页　新闻　贴吧　知道　音乐　图片　视频　地图　文库　更多»

百度为您找到相关结果约1,690,000个　　　　　　　　　　▽搜索工具

《企业会计准则-应用指南》
2014年7月28日 - 财会[2006]18号《企业会计准则-应用指南》附件:会计科目和主要账务处理 会计科目和主要账务处理涵盖了各类企业的各种交易或事项,是以企业会计准则中涵...
www.360doc.com/content... ▼ - 百度快照 - 796条评价

财政部关于印发《企业会计准则——应用指南》的通知

2006年10月30日 - 国务院有关部委,有关直属机构,各省、自治区、直辖市、计划单列市财政厅(局),新疆生产建设兵团财务局,有关中央管理企业:根据《企业会计准则——基本...
www.chinaacc.com/new/6... ▼ 百度快照 - 1192条评价

企业会计准则应用指南——会计科目和主要账务处理_百度文库

★★★★★ 评分:4.5/5 182页
2013年5月8日 - 企业会计准则应用指南——会计科目和主要账务处理 顺序号 编号 会计科目名称 一、资产类 会计科目适用范围说明 1 2 3 4 5 6 7 8 9 10 11 12 13 14 15 ...
wenku.baidu.com/link?... ▼ 评价

企业会计准则应用指南(2006正式版).doc	评分:4/5	154页
最新企业会计准则-应用指南.pdf	评分:4.5/5	398页
新企业会计准则应用指南.pdf	评分:4.5/5	133页

更多文库相关文档>>

《企业会计准则、应用指南、解释等(201610止)》.pdf文档全文免费...

2015年10月10日 - 企业会计准则 截至2014-10-29止版本 Felix Zhu 根据网络公开文档整理
2014-10-31 企业会计准则——基本准则 (2006年2月15日财政部令第33号公布...
max.book118.com/html/2... ▼ 百度快照 - 267条评价

企业会计准则应用指南_百度百科

　　《企业会计准则应用指南》是2007年中国时代经济出版社出版的图书

Baidu百度 企业会计准则解释　　　　　📷　百度一下

网页　新闻　贴吧　知道　音乐　图片　视频　地图　文库　更多»

百度为您找到相关结果约1,010,000个　　　　　▽搜索工具

企业会计准则解释汇总_东奥会计在线

2014年8月8日 - 关于印发企业会计准则解释第6号的通知责任编辑:绿梦分享: 纠错收藏打印上一篇新闻:关于印发《企业会计准则——应用指南》的通知 下一篇新闻: 企业会...
www.dongao.com/news/kj... ▼ 百度快照 - 373条评价

关于印发《企业会计准则解释第7号》的通知

2015年11月4日 - 为了深入贯彻实施企业会计准则,解决执行中出现的问题,同时,实现企业会计准则持续趋同和等效,我部制定了《企业会计准则解释第7号》,现予印发,请遵照执...
www.mof.gov.cn/pub/kjs... ▼ 百度快照 - 评价

新企业会计准则解释_百度文库

★★★★☆ 评分:4/5 39页
2012年5月17日 - 新企业会计准则解释_PPT制作技巧_实用文档。新企业会计准则解释 新企业会计准则解释.txt 鲜花往往不属于赏花的人,而属于牛粪。.道德常常能弥补智慧的...
wenku.baidu.com/link?u... ▼ 百度快照 - 评价

财政部印发企业会计准则解释第8号_中华会计网校

2016年1月5日 - 近日,财政部印发了《企业会计准则解释第8号》,对会计实务中关于商业银行理财产品的一些问题作了明确。一、商业银行应当按照《企业会计准则第33号——...
www.chinaacc.com/tansu... ▼ 百度快照 - 1192条评价

企业会计准则解释

2012年12月11日 - 为了进一步贯彻实施企业会计准则,根据企业会计准则执行情况和有关问题,针对上市公司今年以来执行新会计准则的情况,财政部制定了《企业会计准则解释第1...

例 5.93：

原文：State-owned Land Use Right Grant Contract (Jing Di Chu [He] Zi (03) No. 381) dated 22 April 2010

译文：2010 年 4 月 22 日的《国有土地使用权出让合同》（京地出【合】字（03）第 381 号）

分析：本例的文件名以及文件编号，一看就知道是来自中国，像这种文件，要译成正确的中文，就必须靠猜测加网络搜索验证了。其中 State-owned Land Use Right Grant Contract 的译文"国有土地使用权出让合同"中，唯一不能确定的是 grant 的译文"出让"是否正确。如果将"国有土地使用权出让合同"放在百度里查，可以看到较多的"国有建设用地使用权出让合同"，所以，基本可以确定 grant 译成"出让"是对的。Jing Di Chu [He] Zi (03) No. 381 的译文"京地出【合】字（03）第 381 号"，如果直接放到百度上，可以发现挺多类似的文件编号，所以，可以确定文件编号的译法是对的。

第五节　原文纠错内容

实用翻译过程中，有时会碰到一些因无逻辑或不完整而难以理解的原文，这些问题通常会是原文出现错误导致的。作为一名合格的译员，应该能够识别出尽可能多的原文错误，加以纠正，并以纠正后的原文译出。常见的原文错误有因中文拼音或五笔打字出错的错别字、文字位置错误、文字转化错误、文字格式错误（包括数字上下标）、拼写出错以及中英文混杂使用、原文表达不当等。具体请看以下例子。

一、英文原文纠错

例 5.94：

原译：

| Rated power | 23MVA ONAN , 30MVA |
| 额定功率 | 23 MVA 油浸自冷，30MVA |

ONAF
油浸风冷

Rated incoming voltage ± 2x2.5%　　11000 V
额定输入电压 ± 2x2.5%

Rated outgoing voltage　　132000 V
额定输出电压

Tap Changer　　OCTC –5 to +5% at a step of
抽头转换开关　　OCTC–5 至 +5% 一步
2.5%

Standard　　IEC 60076
标准

Cooling　　ONAN
冷却：　　油浸自冷

Volt Impedance at 30MVA base load　　> 8%
30MVA 基本负荷伏阻抗

改译：

Rated power　　23MVA ONAN , 30MVA
额定功率　　23 MVA 油浸自冷，30MVA

ONAF
油浸风冷

Rated incoming voltage ± 2x2.5%　　11000 V
额定输入电压 ± 2x2.5%

Rated outgoing voltage　　132000 V
额定输出电压

Tap Changer　　OCTC –5 to +5% at a step of 2.5%
抽头转换开关　　无励磁分接开关（OCTC），–5 至 +5%，每级 2.5%

Standard 标准	IEC 60076
Cooling 冷却：	ONAN 油浸自冷
Volt Impedance at 30MVA base load 30MVA 基本负荷伏阻抗	> 8%

分析：这里重点分析 OCTC –5 to +5% at a step of。这一个短句明显不完整，有经验的译员，应该会猜测到 of 后面缺个名词，而下一行的左边却有一个孤立的"2.5%"，所以可以合理怀疑原文把 2.5% 移位到下一格了。所以完整的句子应该是 OCTC –5 to +5% at a step of 2.5%。

这时，翻译的重点又转到 OCTC，而左边有一个 Tap Changer，所以可以认为 OCTC 里的 TC 就是 tap changer，在百度里查 "OCTC tap changer"，应该不难查到其完整拼写是 off-circuit tap changer，意思是"无励磁分接开关"。

那 OCTC –5 to +5% at a step of 2.5% 的意思呢？有点理工背景的，或者想象力丰富一点的译员，就可以猜到其意思是上文修改后的译文。

例 5.95：
原文：

项目		指标
水分，%		≤ 10.0
表观黏度 mPa.s	蒸馏水	≥ 25
	饱和盐水	≥ 15
	搬土浆（180℃ ×16h）	≥ 25
切动力 YP (Pa)	蒸馏水	≥ 9.6

项目		指标
水份，%		≤ 10.0
表观黏度 mPa.s	蒸馏水	≥ 45
	4% 盐水	≥ 5.0
	搬土浆（180℃ ×16h）	≥ 25
切动力 YP (Pa)	蒸馏水	≥ 20
静切力 G10s/G10min (Pa)	蒸馏水	≥ 10 / ≥ 20
漏失量	搬土浆	≤ 30.0

原译：

Item		Index
Moisture, %		≤ 10.0
Apparent viscosity, mPa.s	Distilled water	≥ 25
	Saturated saline water	≥ 15
	Bentonite slurry (180℃ ×16h)	≥ 25
Shearing force YP (Pa)	Distilled water	≥ 9.6

Item		Index
Moisture, %		≤ 10.0
Apparent viscosity, mPa.s	Distilled water	≥ 45
	4% saline water	≥ 5.0
	Bentonite slurry (180℃ ×16h)	≥ 25
Shearing force YP (Pa)	Distilled water	≥ 20
Static cutting force G10s/G10min (Pa)	Distilled water	≥ 10 / ≥ 20
Leakage loss	Bentonite slurry	≤ 30.0

改译：

Item		Index
Moisture, %		≤ 10.0
Apparent viscosity, mPa.s	Distilled water	≥ 25
	Saturated saline water	≥ 15
	Bentonite slurry (180℃ × 16h)	≥ 25
Yield point YP (Pa)	Distilled water	≥ 9.6

Item		Index
Moisture, %		≤ 10.0
Apparent viscosity, mPa.s	Distilled water	≥ 45
	4% saline water	≥ 5.0
	Bentonite slurry (180℃ × 16h)	≥ 25
Yield Point YP (Pa)	Distilled water	≥ 20
Gel strength G10s/G10min (Pa)	Distilled water	≥ 10 / ≥ 20
Leakage loss	Bentonite slurry	≤ 30.0

分析：这里的"切动力"，一开始时不理解是什么意思，在网络上查"切动力"，查不到准确解释。但后来发现原文里还有"静切力"，所以怀疑这个"切动力"是"动切力"的笔误。这时，再到网络上查"动切力"，结果发现百度百科里对"动切力"的介绍里，也有提到静切力，而且还有其英文 yield point（屈服点）。

也就是说，"动切力"实际上就是"屈服点"，专业译法是 yield point，这就是"动切力"后有"YP"字样的原因。这时，我们应该再多个心眼，考虑一下静切力是不是也有既定的专业译法，查一下百度百科，果然，"静切力"的专业译法是 gel strength，实际意思就是"凝胶强度"。所以，这时要把全文的"切动力"和"静切力"的译法都一一修改过来。

二、中文原文纠错

例 5.96：

原文：中交西安筑路机械有限公司

地址：西安市经济开发区泾渭工业园泾商南路

原译：CCCC Xi'an Road Construction Machinery Co., Ltd.

Address: Jingshang South Road, Jingwei Industrial Park, Xi'an Economic Development Zone

改译：CCCC Xi'an Road Construction Machinery Co., Ltd.

Address: Jingshang South Road, Jingwei Industrial Park, Xi'an Economic Development Zone

分析：查"泾渭工业园泾商南路"，查不到相应地址，但查"中交西安筑路机械有限公司 泾渭工业园"，可以查到是"泾高南路"，所以判定为"高"被误写为"商"。

例 5.97：

原文：具体要求：灭菌包外应有化学指示物，高度危险性物品包内应<u>旋转</u>包内化学指示物，置于最难灭菌的部位。

译文：Specific requirements: chemical indicators outside sterile pack are needed, and chemical indicators should be put into the place that are hardest to sterilize of the pack of high harmful substances.

分析：本例的"旋转"一词，跟上下文很不协调，因此怀疑为打字错误。这里推测为五笔打字错误，"旋转"应该是"放置"的笔误（打五笔都是 ytlf），这从后文"置于"一词也可以证实。

也就是说，在做翻译时，要充分展开各种合理联想。因为翻译理论中就有一个"**关联理论**"，通常而言，最佳关联就是合理译法或者合理纠错法。

例 5.98：

原文：区域内由北东向南西分布有四条大断裂 F1、F2、F3、F4，走向 3000-3250。

原译：The area has four faults F1, F2, F3 and F4 From NE to SW respectively, with strike of 3000-3250.

改译：The area has four faults F1, F2, F3 and F4 From NE to SW respectively, with strike of 300º-325º.

分析：有地质背景知识的人，肯定会对这里的"走向3000-3250"的表达感觉很奇怪。经合理推测，这应该分别是"走向3000-3250"是"走向300º-325º"的笔误。

例 5.99：

原文：The operation of the crane shall be completely electrical with 415 Volts, 3 phases, 50 cycles/Sec AC.

原译：起重机应完全由415伏特，3相，50周期/秒的交流电力操作。

改译：起重机作业应完全使用415V、3相、50Hz的交流电。

分析：原文中cycles/sec的字面意思是"周期/秒"，但在电学中，其意思实际上就是频率单位"赫兹（Hz）"，所以，应该将其纠正改译成Hz。

本节课后练习

翻译下列句子，注意下画线部分：

1. For instance, when a source located at grade generated 90 dB of pressure level at 10m from the source, the following sound power level is obtained with equation (2) (5 being the surface area of a hemisphere with a 10m radius):

   Lw = Lp + 10 log10 S + A

   　　= 90 + 10 log10 (2 π x 102) + 0

   　　= 90 + 10 log10 (628) = 118 dB (A)

2. The side-looking antenna enables the same 5-in. (12cm) outside diameter (OD) tool to be used in all hole sizes larger than 5⅝-in. (149 mm).

3. 红色灯光表示越限报警或紧急状态：黄色灯光表示预报；绿色灯光表示运转设备或过程变量正常。

4. 要有足够的胎体强度和硬度，避免受冲击作用而破裂，一般选用中硬2硬胎体，唇面形状以平底形、圆弧形和同心圆尖齿形为好。

第六节 专有名词的译法

专有名词包括节日名称，也包括人名、地名、组织机构名、职务职称等。这些名词基本都有既定译法，其中人名、地名、组织机构名、职务职称等，除了自己能够记住其译法的，都必须通过网络搜索等方法查找，才能确定其译法。本节希望通过一些典型专有名词的译名确定过程，让译员掌握一些基本的专有名词网络搜索法，并由专有名词的译法，引申介绍汉字文化圈的文化相通性。

一、专有名词译法的确定

（一）知名人士姓名的确定

例 5.100：

原文："5G mobile networks can provide a true-to-life video communications experience," said Eric Xu, rotating chief executive officer of Huawei.

原译：华为轮值 CEO <u>埃里克·徐（Eric Xu）</u>表示："5G 移动网络可以实现逼真的视频通讯体验"。

改译：华为轮值 CEO <u>徐直军</u>表示："5G 移动网络可以实现逼真的视频通讯体验"。

分析：本句的 Eric Xu，一般译员都无法确定其具体中文名。但是，作为一个合格的译员，应该想到，这个 Eric Xu 在华为公司（中国最具实力的公司，世界 500 强）应该是个名人，而且是个中国人，因此肯定有中文名。所以，可以到网络上查找，果不其然，不论是在百度还是必应（Bing）里，查找"华为 Eric Xu"，都可以知道其中文名为"徐直军"。

（二）地名、道路名

例 5.101：

原文：孟加拉国<u>古拉绍</u>300～450MW联合循环燃气电厂工程

原译：Bangladesh <u>Gulashao</u> 300~450MW Combined Cycle Gas-Fired Power Plant Project

改译：Bangladesh <u>Ghorasal</u> 300~450MW Combined Cycle Gas-Fired Power Plant Project

分析：孟加拉国的地名"古拉绍"太有名，所以在网络上不容易查到其英文。

对于这种地名，只能充分利用上下文，在网络上搜索其英文名。在本例中，可以直接把原译文除错误译名 Gulashao 之外的相关内容放在网络上搜索。比如，考虑到该工程在孟加拉国是个大工程，在百度上搜索 "Bangladesh 300~450MW Combined"，结果就可以直接找到跟中文"古拉绍"发音类似的地名 Ghorasal 紧跟在 Bangladesh 之后，可以推断这个词就是"古拉绍"的英文译名。

(三) 组织机构名

例 5.102:

原文：湖南中车时代电动汽车股份有限公司

原译：Hunan Zhongche Era Electric Vehicle Co., Ltd.

改译：Hunan CRRC Times Electric Vehicle Co., Ltd.

分析：这个公司名，到百度里查公司官网的链接并点开，就可以直接找到英文名：

把公司名直接放到百度或 Google 查官网，点击官网链接后，有正式英文名的就可以找到了。

例 5.103：

原文：Toepfer International – Asia Pte. Ltd.

100 Beach Road #26-01 Shaw Tower, Singapore, 189702

原译：托福国际亚洲有限公司

（Toepfer International – Asia Pte. Ltd.）

新加坡海滩路 100 号萧氏大厦 #26-01

改译：托福国际亚洲有限公司

（Toepfer International – Asia Pte. Ltd.）

新加坡海滩路 100 号邵氏大厦 #26-01

分析：这里的 Shaw Tower，感觉上可能是以某个名人的名字命名的大厦，但译成"萧氏大厦"是否准确还必须查证一下。通过在百度里查"100 Beach Road #26-01 Shaw Tower, Singapore, 189702 新加坡 大厦"，好像不容易查到相应的大厦名称，但查"Shaw Tower, Singapore 新加坡大厦"，就可以发现其名称为"邵氏大厦"，而且还可以了解到，这个大厦跟邵氏电影王国有关，所以，可以断定，这个邵氏，就是邵逸夫，而且，这个邵逸夫给中国很多高校里都捐献了图书馆，其中很多都起名为"逸夫楼"。

例 5.104：

原文：Shenzhen Ping'an Financial Technology Consulting Ltd.

Shenzhen Joint Financial Data Technology

原译：深圳平安金融技术咨询有限公司

深圳联合财务数据技术有限公司

改译：深圳平安金融科技咨询有限公司

深圳联合金融数据科技有限公司

分析：在将原来是中国公司的英文名回译成中文时，通常不能根据自己的想法译成中文，否则会译错。一个较好的办法是在百度里以自己想定的译名搜索相应公司名，这时输入框里会自动跳出实际中文名。当然，在百度里以英文名加上可以确定的中文关键词一起搜索也是一个办法，如在百度里搜索"Shenzhen Ping'an Financial Technology Consulting Ltd. 深圳平安"，就可以迅速找到这个公司的正确中文名。

但另一个公司名可能是因为不够出名，用这种方法搜索不到。

像这样的例子，我们还可以找到很多，这里不多举例。总之，在涉及一些知名大公司的主管姓名的翻译时，如果是中国人名中译英，要到网络上查找一下该主管有无惯用的英文名，如有，必须使用该英文名；如果是外国人名中译英，一般都要确认一下其正确的外文拼写法。如果是中国人名英译中，通常就要到网络上查一下这个人名的正确中文名称；如果是外国人英译中，一般只要音译（但第一次译出时要在括号后加注英文），或者干脆不译。

二、汉字文化圈文字的相通性

汉字文化圈的覆盖地域主要为现代的东亚和南亚地区，除中国外，还包括在古代主要使用汉字的越南、朝鲜半岛、日本列岛、琉球群岛等；有时候也将现代的美国、加拿大等北美及巴西等南美、马来西亚、新加坡等东南亚的汉族人较多的地区包括在内。其中，朝鲜半岛和越南在很长一段时期内都以汉语为书面语，只是到近代才废除汉字使用本族语。如今，越南语、朝鲜语和日语三语的书写字仍有大量源于汉字。

汉字文化圈的很多词语，尤其是人、地、组织机构的名称，通常都有中文名，或者中文和外文同用。因此，这些地方的专有名词的英文，通常都能找到中文痕迹。比如，美国的旧金山（San Fransisco）、檀香山（Honolulu）、印尼爪哇省省会三宝垄（Semarang，三宝为郑和小名）、马来西亚柔佛州首府新山（Johore Bahru，即柔佛巴鲁）等。

此外，中国大陆有些地区的方言，如闽南语和粤语，也有自己独特的发音体系，甚至有自己的文字，所以其词语的英文有时也会跟正常的汉语拼音有所不同，但也有中文痕迹。比如，孙中山（Sun Yat-sen，即孙逸仙）、李小龙（Bruce Li）等。

所以，当将这些国家地区的名词，尤其是专有名词译成汉语时，就可以根据其汉语痕迹，帮助译者推测出其大概意思，并通过网络搜索进一步确定。

以上两种情况，一般都要把英文译成确定的中文名，不能不译。而且，这些人、地址和组织机构的名称译成中文后，通常后面不必再保留原英文，因为中文就是这些名词的源语言。

下文列举的，是汉字文化圈一些专有名词中文译名的翻译方法。

例 5.105：

原文：Fernando Chui Sai On will replace Edmund Ho Hau Wah who has governed Macau since the enclave was returned to China in December 1999 after 442 years of Portuguese colonial rule.

译文：崔世安将接替何厚铧出任澳门特首，在1999年12月回归中国之前，澳门曾作为葡萄牙的殖民地由葡萄牙统治442年。

分析：本句的 Fernando Chui Sai On（崔世安）和 Edmund Ho Hau Wah（何厚铧）分别是澳门的第二位和第一位特首，由于澳门同时使用葡萄牙语和粤语，所以澳门人与广东人、香港人一样，其姓名拼音英译时，都用粤语拼音。粤语拼音与威妥玛拼音有点相似，但又有很大差别。

例 5.106：

原文：Flat E, 26th Floor, Fu Kar Yuen, chi Fu Fa Yuen

译文：置富花园富嘉苑 E 座 26 楼

分析：这个地址是香港的地址，拼音是广东话发音。虽然广东话我们不懂，但 Fa Yuen 这个发音，似乎还是可以知道是"花园"，在网络上查"Flat E, 26th Floor, Fu Kar Yuen, chi Fu Fa Yuen 花园"，就可以知道 Fu Kar Yuen, chi Fu Fa Yuen 是"置富花园富嘉苑"，至于 Flat E, 26th Floor 的译法，则不大确定。但译到这个程度已经足够了。

又如：

The Hong Kong and Shanghai Banking Corporation Limited

Address: 1 Shan Mong Road, Kowloon, Hong

汇丰银行

地址：香港九龙深旺道1号

例 5.107：

原文：Ochang-eup, Cheongwon-gu, Cheongju-si, Chungcheongbuk-do, Korea

译文：韩国清忠北道清州市清原群梧仓邑

分析：大家注意一下，这个韩国地名很像汉语拼音或汉语方言发音。这是因为包括韩国和朝鲜在内的朝鲜半岛在几十年前的官方语言还是中文，地名全都是中文。所以朝鲜和韩国的地名英文很像汉语拼音。此外，越南也是几十年前才开始废弃汉字，因此，越南的地名也很像汉语拼音。而且,这些国家的人,有很多也是直接取汉语名,然后再配以本国文字(实际上就是拼音文字)。

例 5.108:

原文:VN-Ministry of Public Security

VN-General Police Department for Crime Prevention and Suppression

VN-Supreme People's Court

原译:越南——公安部

越南——犯罪预防和抑制警察总局

越南——最高人民法院

改译:越南——公安部

越南——越南公安部预防和打击犯罪警察总局

越南——最高人民法院

分析:本例是对越南国家机构英文名的汉译,其中第二个比较难译。由于一般译员都不懂越南语,因此无法知道第二个机构中文名是什么。但由于越南是原汉字文化圈国家,在几十年前还完全使用汉字,因此,其组织机构名通常都会有确定的汉字译法。但怎么查呢?

具体查法是,直接抛弃英文,在百度上查相关关键词,如"越南犯罪预防与",结果如下:

打开第二个链接"越南公安部 百度百科",其中有一处内容为"人民公安警察总局(原名预防和打击犯罪总局,第六总局):总局长范文永中将、副总局长杜金线。"而"预防与打击犯罪总局"的译法与

General Police Department for Crime Prevention and Suppression 的意思非常贴近，再到百度上查"越南预防与打击犯罪总局"，可以发现"越南预防与打击犯罪警察总局"的提法较多，因此，将其译成"越南公安部预防和打击犯罪警察总局"。

本节课后练习

翻译下列句子，注意下画线部分：

1. Yang Yuanqing said Lenovo was also sharpening its focus on artificial intelligence (AI), but different from the strategy of Chinese internet giants Baidu, Tencent Holdings and Alibaba Group Holding, the parent company of the South China Morning Post.
2. 公路是从矿区起点沿巴波卢（Bapuluo）至蒙罗维亚（Monrovia）国道向南2.9km，再折向西南30km，跨越圣保罗河后，连接至邦矿铁路的邦矿站，全长32.9km。
3. 甲　　方：绿地（英国）投资有限公司

   联系地址：伦敦贝森浩街40号11楼

   联系电话：＊＊＊＊＊＊＊＊＊＊＊

   乙　　方：上海麦阁文化发展有限公司

   联系地址：上海市徐汇区华亭路99弄1号

   联系电话：＊＊＊＊＊＊＊＊＊＊＊
4.
   > **TO**: ExxonMobil Asia Pacific Research & Development Co., Ltd.
   > 　　17th Floor Metro Tower, 30 Tian Yao Qiao Road
   > 　　200030 Shanghai, P.R. China
   >
   > **From**: Nantong No.4 Construction Group Ltd.
   > 　　31#438 Lane, Ningguo Road,
   > 　　Shanghai 200090, P.R. China.

5. Headquarters: LG Twin Towers, 20 Yeouido-dong, Yeongdeungpo District, Seoul 150-721
6. 马来西亚雪兰莪州巴生港口海南村

第七节 其他

一、非专业用词的翻译

在科技类文章里，除了专业用词有基本固定的专业译法之外，有些非专业用词，特别是一些通用的形容词、副词等，在中英互译过程中，也有基本对应的译法。具体请看以下例子。

例 5.109：

excellent、strong、presence、good、total、substantially

原文：In addition, OMS-2 has been found to possess excellent hydrophobicity and strong affinity toward organic compounds, capable of selectively adsorbing volatile organic compounds in the presence of water vapor, thus retaining good durability in total oxidation of hydrocarbons where water is formed via oxidation reactions.

原译：此外还发现 OMS-2 具有极佳的疏水性（hydrophobicity），而且对于有机化合物具有极强的亲和力，能够选择性地吸收以水汽形式呈现的挥发性有机化合物，在这个过程中出现的水是通过氧化反应形成的，这样就保持了在碳氢化合物的整个氧化过程中的良好的耐久性。

改译：此外，还发现八面体分子筛 OMS-2 具有优异的疏水性，并对有机化合物具有强的亲和力，能选择性吸收在水蒸气中存在的挥发性有机化合物，因此在生成水的烃类完全氧化反应中保持较好的耐用性。

分析：这一个句子，不但要注意专业用词的翻译准确性，还要注意非专业用词的翻译准确性，如 excellent, strong, presence, good, total，其中：

excellent 是优异，译成极佳也差不多，

strong 是强，译成极强就有点过分，

presence 是存在的意思，在化学中是很常见的用词，

good 译成较好或良好都可以，

total oxidation 则是完全氧化，而不是整个氧化过程。

另外，这一句最后部分 where water is formed via oxidation reactions 看似不起眼，实际是一个重要的限定性状语，表示前面的 total oxidation 是什么样的一个过程，跟前面的 hydrophobicity（疏水性）形成呼应，要不就没有必要说 OMS-2 有疏水性了，这是本句逻辑的关键所在。

例 5.110：

low、lower

原文：Exhaust air from paint shops running E-Cube with air recirculation only contains low concentrations of particulate. As a result, VOCs can easily be treated by adsorption systems plus thermal oxidizers. This creates an ideal basis for achieving VOC emissions thresholds and the lower permitted values due to be introduced in China for certain substances.

原译：安装 E-Cube 系统（附带空气再循环系统）后，涂装车间排出的空气的颗粒浓度极低。因此，只需吸附系统和热氧化器，即可轻松处理挥发性有机化合物，满足挥发性有机化合物排放限值和其他待引进中国之特定物质低值标准。

改译：安装（附带空气再循环系统的）E-Cube 系统后，涂装车间排出的空气只含有低浓度的颗粒物。因此，挥发性有机化合物可被吸附系统和焚烧炉简单处理，这为即将引进中国的特定物质的挥发性有机化合物的排放阈值以及更低允许值创造了理想的条件。

分析：本例的 low concentrations 直接译成"低浓度"即可，如译成"极低浓度"就是不准确翻译，后面的 lower permitted values 则是"更低允许值"，译成"低值标准"也不准确。

从以上 excellent、strong、good、low、lower 等词的译法可知，在翻译这些表示程度的词时，一般只要按其字面意思译出，如果是原形词，如 strong、good，只需译成"强的""好的"，个别情况下要在前面加个"较"字（如：较强[的]、较好[的]），如果是比较级，则在前面加个"较""更"等词。如果在前面随便加上"极""很""非常"，则很可能导致夸张表达。

例 5.111：

significant

原文：Compared with traditional models, *i*Recovery® system delivers significant reductions in water consumption, sulfuric acid condensation and thermal shock, thereby extending plant duration and cutting maintenance requirements.

原译：与传统的模型相比，*i*Recovery® 系统明显减少了耗水量、硫酸冷凝和热冲击，从而延长了设备寿命，降低了维修量。

改译：与传统的模型相比，iRecovery® 系统显著减少了耗水量、硫酸冷凝和热冲击，从而延长了设备寿命，降低了维修量。

分析：本句的 significant 译成"明显（的）"也没错，但在科学技术类文件里，significant 更常见的译法是"显著（的）"。在统计学中，significant 更是几乎全部译成"显著（的）"，如 significant effect（显著性影响）。

例 5.112：

measured

原文：Regain permeability measurements were requested, but the maximum measured gas permeability on this sample set was less than 0.03 millidarcies, which is too low for regained permeability testing.

原译：这里要求进行恢复渗透率测量，但该样品组所测量的最大渗透率低于 0.03md（毫达西），这样的渗透率对于恢复渗透率试验来说太低。

改译：这里要求进行恢复渗透率测量，但该样品组的最大实测渗透率低于 0.03md（毫达西），这样的渗透率对于恢复渗透率试验来说太低。

分析：本例想说明的一点是，measured 在很多情况下，都是"实测（的）"的意思，如果译成"（所）测量（的）"就流于字面翻译了。

例 5.113：

typical

原文：Typical errors between calculated and measured ECD were ~1%.

原译：计算与实测当量循环密度之间典型误差约为 1%。

改译：计算与测量当量循环密度之间一般误差约为 1%。

分析：本例想说明的一点是，不要一碰到 typical 就理解成"典型（的）"，其实在应用类翻译中，在很多情况下，typical 都是"一般（的）""通常（的）"的意思。在此，大家可以体会一下，"典型"和"一般"在中文中是不是有些时候意思是相近的。

例 5.114：

substantially

原文：The recommendations of the Operation Assessment that are relevant to the resumption of production of the oilfield are substantially implemented, and a detailed corrective action plan has been developed for those that are still progressing.

译文：与该油田恢复生产相关的经营评估建议基本得到了执行，而且为那些仍在进行的建议制订了详细的矫正行动方案。

分析：substantial/substantially 经常在句子中表示"基本（上）""大体（上）""实质（性、上）""几乎 / 完全"的意思。

又如：

例 5.115：

原文：As shown below, theoretical product transition time can be substantially shorter than that required for a continuous stirred tank reactor (CSTR) or fluidized bed gas phase process.

译文：如下图所示，理论产品过渡时间明显短于连续搅拌槽式反应器（CSTR）或流化床气相工艺所需的产品过渡时间。

分析：本句的 substantially 是"大大（地）""明显（地）"的意思。

例 5.116：

proposed

原文：Users of the evaluation report are the Client and all relevant parties involved in the proposed project.

原译：评估报告的使用者为委托方及涉及提议工程的所有相关方。

改译：评估报告的使用者为委托方及涉及拟定工程的所有相关方。

分析：proposed：拟定的、拟议的。又如 activities proposed to be taken：拟采取的行动。

二、非正常词义

例 5.117：

deliver

原文：Guidelines referred to in point (a) of paragraph 1 shall be drawn up by the date referred to in the second subparagraph of Article 28 and shall be regularly updated, taking into account the recommendations delivered in accordance with Article 5 as well as scientific and technical progress.

译文：第1小段第（a）点提到的指南必须在第28条第2款所述的日期前拟定，且必须定期更新，同时将根据第五条提交的建议和科技发展考虑在内。

分析：本例中的难点是 deliver 意思的确定。

查中国译典，deliver 有"供给"的意思，但实际上，我们还可以译成"提供"。

也就是说，有时候针对个别单词，我们不必拘泥于字典里的译法，而是根据上下文，采用合适的译法。而这种译法，如果在英语里经常出现，说不定词典里就会另外添加这样的一种解释。也许，这就是中外交流中单词译法慢慢丰富的一个过程，也就是说，可能某个单词最开始时，在词典里只有五种中文意思，而后来中国人慢慢发现其实这个词有第六种意思，就在词典里添加这种意思。当然，这种意思一般不会非常常见，因此，这个新增意思通常会被放在该词条的最后。

三、原文同位语或解释性表达的翻译

例 5.118：

原文：By better understanding the causes of health and disease and making best use of big data, we can develop better diagnostics (e.g. in vivo medical imaging), therapies (e.g. clinical trials for non-communicable diseases), health promotion and disease prevention strategies (e.g. environment and health based interventions) at the personal and population levels, as well as technologies to support healthy ageing (e.g. mHealth applications) and in poverty related and antibiotic resistant infectious diseases (e.g. vaccine platforms for HIV/AIDS and tuberculosis) with particular emphasis on a triangular relationship involving China, Portugal and Portuguese speaking African countries.

译文：通过更好地了解促进健康和引发疾病的原因，及更加充分地利用大数据，在人和群体层面，我们可以研发出更好的诊断法（如体内医疗成像）、治疗法（如非传染性疾病的临床试验）、健康促进和疾病预防策略（如基于环境和卫生的干预措施），促进健康老龄化的支持技术（如移动医疗应用程序［mHealth］），及针对由贫困造成的、对抗生素有抗药性的感染性疾病（如针对艾滋病和结核病的疫苗平台）的技术，其中特别强调了涉及中国、葡萄牙和非洲葡语国家的三方关系。

分析：本段中的最后一个括号里的 e.g. vaccine platforms for HIV/AIDS and tuberculosis，应该只是把一个意思（艾滋病）的两个表达法都列出来，并不是表达两个意思。

例 5.119：

原文：Assuring that the hotel is well represented by employees trained in C.P.R.—cardiopulmonary resuscitation—and the Heimlich (choke prevention) manoeuvre.

原译：确保酒店受训员工在心肺复苏（CPR）。——心肺复苏术——和哈姆立克急救法（防止窒息）方面表现优秀。

改译：确保酒店受训员工在心肺复苏术和哈姆立克急救法（防止窒息）方面表现优秀。

分析：注意，原文"C.P.R.—cardiopulmonary resuscitation"里的 C.P.R. 就是破折号后内容的缩写，所以只要译一次就可以，而且译文可以不用破折号。

四、英文名词的单复数

中译英译到名词时，要考虑一下是单数还是复数，感觉确定是复数的，就用复数格式，如果是单数的，通常要考虑到加冠词。

例 5.120：

原文：借助历史地图，圈画出各大洲的文明发祥地

原译：By virtue of historical atlas, circle birthplace of civilization of each continent

改译：Circle the birthplaces of civilization of all continents with historical atlas

分析："各大洲"是指所有大洲，要译成 all continents，发祥地不止一个，所以要用复数。

例 5.121：

原文：将锅炉口的螺母旋紧，然后通电。

译文：Tighten all bolts of the furnace, and turn on the power.

分析：锅炉口的螺母，都不可能只有一个，所以这里将"螺母"译成 all bolts。

例 5.122：

原文：箱/盘的固定方式应一致，配电箱/盘的面漆颜色应统一、二、三级配电箱/盘应装设漏电保护自动空气开关，漏电保护应按一机一闸配置，所有配电箱/盘应设独立的接零和接地母排。

原译：Box/plate fixation should be consistent, the paint color of distribution box/panel should be uniform. Second and third level distribution box/panel should be equipped with leakage-protection automatic air switch. Leakage-protection device should be equipped per machine per switch. All distributor box/plate should be designed with PE or PEN Busbar.

改译：Box/panels should be fixed in consistent way, and the topcoat of distribution boxes/panels should be in uniform color. Secondary and tertiary distribution boxes/panels should be equipped with automatic leakage-protection air switches, with one air switch provided to one equipment, and all distributor boxes/panels should have one independent PE or PEN busbars.

分析：本例原译中除了其他问题之外，还把所有名词都译成单数。实际上，在做中译英时，通常可以根据上下文或者逻辑判断名词的数，从而确定名词是否用复数形式以及谓语是否用第三人称单数。

引申：

应用翻译中的名词单复数的确定，很难有一个普遍适用的标准。但应特别注意的是，某些名词，如机械（电子、电气）设备中的螺丝、螺母、螺栓、齿轮的齿、导线、管道、接头（接点）、电极、磁极、阀门、龙头等；几何中的棱、角、边；生物（医学）中的枝、叶、细菌、血管、细胞等，这些通常都不会是只有一个的，所以译成英文的时候，通常可以译成复数。但若明确地知道这些名词所示的数只有一个时，就必须用单数（第一次出现时前面加 a/an）。当然，很多时候，中文原文的名词是单数还是复数，需要译者通过自己掌握的常识加以推理，做出理性判断。对于其他的词，通常也可以根据上下文、常识、逻辑等判定其单复数。

第六章　其他相关技巧与知识

第一节　各种文字的辨识

在翻译过程中，有时会碰见 PDF 格式或图片格式的原文，里面出现无法转化且难以辨认的繁体字（又称正体字）、异体字、手写字或艺术字体。以下分别举例分析（不翻译，只认字）。

一、繁体字与异体字

例 6.1：

原文：ABC 公司年终彙报

分析：本例的原文是扫描图片，文字无法复制到网络里辨认。原文其中有一个字"彙"很多人都不认识，这是"汇"的繁体字。简体字"汇"对应至少三个繁体字，其中"匯"的使用场合为"滙合""滙率""滙款""滙流"等，"滙"等同于"匯"，而"彙"的使用场合则是"词彙""彙報""彙集""彙編"等。

如果 PDF 或图片原文里不认识的繁体字或异体字无法辩认，而且不能通过上下文推测其所对应的简体字，译员就只能求助于他人了。

其实，作为译员，如果想长期从译，就很有必要学习繁体字和异体字，学习办法很多，包括在网络上查找并学习简繁体字对照表和异体字表，以及在网络上看繁体字的文章等。这里，推荐大家上网查找一位中国台湾学者的介绍性文章《簡化字、異體字與正體字對照表》。

二、一字读两音

汉字里一字双音节的现象,基本上是计量单位,其中很多词初次碰到的人会看不懂,或者很容易被误解。这些词现在已经极少使用,但还是偶尔会碰到,现列举如下(圆括号内为读音):

[瓩](十克)、[瓩](千克)、[瓱](毫克)、[瓰](分克)、[瓸](百克)、[甅](厘克)、[嗧](加仑)、[瓩](千瓦)、[粏](厘米)、[粁](千米)、

[粍](毫米)、[糎、浬](海里)、[哩](英里)、[嗘、浔、口寻](英寻)、[呎](英尺)、[吋](英寸)

[竔、矸](公升)、[唡](蒲式耳)、[呏、嗧](加仑)

三、手写字

例 6.2:

原文:户口所在地 <u>安徽</u> 省(市) <u>巢湖无海</u> 区(县) <u>石涧镇</u> 街道(乡镇)

分析:对于本例,由于译员不是安徽人,因此无法定"巢湖"后面的两个字是什么字。在网络上查"巢湖行政区划",一时也查不到,但查"巢湖有几个区"或"巢湖县石涧镇"就可以查到了,原来这是"无为"县。列举本例的目的是想让译员知道,碰到无法辨认的手写字(或扫描字)地址时,通常可以通过网络查找推知手写字(或扫描字)是什么字,而不必求助他人。

至于其他非地址类手写字,就需要译员自己懂得辨认或者让他人帮忙辨认,实在认不出时,可以留空不译,但更好的做法是用 ××× 或 *** 等代替看不清楚的字。

四、艺术字

例 6.3:

原文:

分析:本例中的白色英文艺术体字母,很难辨认。但对于认识一些常见英文艺术体写法来说,并不难辨认出这是 comfortable。

本例的目的,是想提醒译员有意识地学会识别英文艺术体或其他非正常字体的文字。

五、不完整(截断或模糊)文字的猜测

例 6.4:(猜测最下方的不完整文字)

原文:

DESCRIPTION OF GOODS	:	1,600 Cans of Glyphosate.
BILL OF LADING NUMBER	:	CNCL472641 issued clean at Shanghai China on 02nd May, 2014.
PLACE/DATE OF SURVEY	:	Kenya Ports Authority Kapenguria area 0n 20th June, 2014. And 02nd July, 2014.
SCOPE OF WORK	:	Survey to determine the condition of the container /the Cargo, extent of damage/loss and ascertain the possible cause.

原译:货物说明:1,600 箱草甘膦

提单号:CNCL472641(2014 年 5 月 2 日于中国上海签发的清洁提单)

调查地点/日期:肯尼亚卡彭古里亚地区港务局,2014年6月20日及2014年7月2日

调查范围:调查并确定集装箱/货物的情况、损坏/损失的范围 *****。

改译:货物说明:1,600 箱草甘膦

提单号:CNCL472641(2014 年 5 月 2 日于中国上海签发的清洁提单)

调查地点/日期:肯尼亚卡彭古里亚地区港务局,2014年6月20日及2014年7月2日

调查范围:调查并确定集装箱/货物的情况、损坏/损失的范围并确定可能的原因。

分析:原文文字不完整,我们推测是 assertain the possible cause。这一例的目的是让译员提高猜词能力,给客户留下更好的印象。

第二节 数字、量级与单位表达法

一、阿拉伯数字的表达法、数字量级

1. 阿拉伯数字表达方式、有效数字的保留

在国际上,阿拉伯数字有好几种常见表达法。这里介绍如下几种:

(1)法国、芬兰、匈牙利、西班牙、葡萄牙、意大利和加拿大法语地区:1 234 567,89。

(2)德国、荷兰、比利时、丹麦、意大利、罗马尼亚和欧洲大多地区:1 234 567,89 或 1.234.567,89;也可能会写成 1·234·567,89,而瑞士德语地区则为:1'234'567,89。

(3)英国、美国、澳大利亚、新西兰、加拿大英语地区、日本、朝鲜和韩国、马来西亚、新加坡:1,234,567.89 或 1,234,567·89。

(4)国际通行写法:1 234 567.89 或 1 234 567,89。

在中国,通用的阿拉伯数字写法,是1234567.89、1 234 567.89 或 1,234,567.89。但作为译员,在碰到其他不同写法时,必须能够理解其意思,并将不符合中国人书写习惯的数字表达法进行转化。但是,需要特别指出的是,1 234 567.89 这样的表达法,如果放在行末时,会导致数字中间断行,所以这个表达法也要慎用。

除此之外,阿拉伯数字的有效数位也要特别注意。比如:原文的3.500 或 1.0,在翻译时一般要照抄,不能认为后面的零是多余的而舍去,因为这些零通常也是有效数位。

2. 数字量级符号

在科技类文章里,一些单位前面会出现G、B、M、k、m、μ、n、p等字母,如:kg、μm、pF,这些字母都是数字量级单位。为帮助各位译员了解这些量级含义,以下列出科技文献里常使用的数字量级的意思和国际符号:

量级	英文	中文	国际符号	量级	英文	中文	国际符号
10^{18}	exa	艾(可萨)	E	10^{-1}	deci	分	d
10^{15}	peta	拍(它)	P	10^{-2}	centi	厘	c
10^{12}	tera	太(拉)	T	10^{-3}	milli	毫	m
10^{9}	giga	吉(咖)	G	10^{-6}	micro	微	μ

（续表）

量级	英文	中文	国际符号	量级	英文	中文	国际符号
10^6	mega	兆	M	10^{-9}	nano	纳（诺）	n
10^3	kilo	千	k	10^{-12}	pico	皮（可）	p
10^2	hecto	百	h	10^{-15}	femto	飞（母托）	f
10^1	deca	十	da	10^{-18}	atto	阿（托）	a

特别提醒：以上量级的国际符号，大小写是有规定的，不能弄错。比如：1MW 是一兆瓦，如果错写为 1mW，就变成一毫瓦了。除了 M 和 m 不能写错外，其他量级的国际符号也不能写错，否则，轻则符号不规范，重则造成意义上的误解。

二、货币单位符号和货币量级

在应用类文件里，经常都会碰到货币符号的缩写，有些常见货币符号的缩写（如 USD、Eur 等）很容易记住，但对于一些不常见的，有些译员甚至可能都不知道是货币符号。所以，以下介绍一些不常见的货币符号的网络搜索方法。

例 6.5：

原文：Company's main financial information from 2008 to 2011

Revenues: R.S.$9.0 million

Gross Profit: R$ 3.9 million

Gross Margin: 43%

原译：2008 年至 2011 年公司的主要财务信息

收入：九百万 R.S. $

毛利润：三百九十万 R.$

毛利率：43%

改译：2008 年至 2011 年公司的主要财务信息

收入：九百万雷亚尔

毛利润：三百九十万雷亚尔

毛利率：43%

分析：在网络上查 "R.S.$ 是什么货币" 或 "R$ 是什么货币"，都找不到相应解释。但查 "R.S. $ is what currency" 或 "R$ is what currency"，

则有很多回复都说明是巴西雷亚尔，而且，本段内容也是与巴西石油贸易相关的文件里的一部分。所以，基本可以确定 R.S.$ 和 R$ 都是巴西雷亚尔。

例 6.6：

原文：Maybe, if somebody has a heart condition, he or she may lodge claims of Rs 2 lakhs, Rs 3 lakhs or even 5 lakhs against the insurer.

原译：也许，如果某人得了心脏病，他或她就会向保险公司发起 Rs 2 lakhs、Rs 3 lakhs 甚至 5 lakhs 的索赔。

改译：也许，如果某人得了心脏病，他或她就会向保险公司发起 20 万、30 万甚至 50 万卢比的索赔。

分析：本例的难点是 Rs 2 lakhs, Rs 3 lakhs or even 5 lakhs。在百度里查 "Rs. 是"，查不到想要的意思。但由于句子中有 lodge claims of ... against ... 这样的表达，感觉好像这是跟保险索赔金额有关的。所以，再查 "Rs. lakhs 是"，可以发现 Rs. lakhs 与印度货币单位有关。再查 "Rs. lakhs 印度"，可以见到有些链接说 Rs. 是印度卢比，lakhs 是货币的 "十万"，也就是说，Rs. 50 lakhs 其实就同 500 万印度卢比，而且这样的意思还可以在网络上得到确认。与 lakhs 类似的词汇还有 crore（千万）。

小结：需要说明的是，当碰到数字前面加字母，而且句子语境与金额相关时，就必须考虑到其与货币的相关性，并有网络上有针对性地搜索，这样，很可能就会查到想要的结果。

三、单位的译法

各种数量和货币金额的一般表达法如下：

（一）物理单位表达法

不论是中文还是英文，物理单位一般都以字母符号表达居多。作为译者，不论原文是用文字还是字母符号表达，建议都将其译成字母符号，只有在单位之前没有数字时才表达成文字。比如：

The standard atmospheric pressure is 101325 pascal.（标准大气压是 101325Pa）

但是 "单位：mm" 和 "unit: mm" 与 "单位：毫米" 和 "unit: millimeter"

这样的表达都是允许的；不过用字母符号表达还是更为普遍的用法。

（二）金额表达法

金额与数字的表达法，一般是货币代号后跟数字，或者数字后跟货币单位，不能在数字前加货币代号，又在数字后加货币单位。如：

  RMB 1234 或 1234 yuan（但 RMB 1234 Yuan 的表达是错的）

另外，货币的大小写表达如下：

  大写：人民币一百元整（小写：RMB100）

  In words: RMB one hundred only（in figures: RMB100）

其中：

大写 in words（即以文字书写）

小写 in figures（即以数字书写）

整 only

金额英译中时，必须用中文说明币种，比如 RMB2340 元译成"人民币 2340 元"，EUR3500 元译成"3500 欧元"。

（三）计量单位的理解与翻译

例 6.7：

原文：The plant shall be provided with compression units with an output > 500,000Sm3/d or any process units where pressure is higher than 70 barg.

原译：工厂应配备产量大于 500000 标方/日 的压缩设备或压力大于 70 巴（表压）的工艺单元。

改译：工厂应配备产量大于 500000 Sm3/d 的压缩设备或压力大于 70barg 的工艺单元。

分析：技术类文件中所出现的单位，如本例中的 Sm3/d（标准立方米/日）和 barg（巴，表压），不管在中文里还是英文里，都应尽量采用符号表示，这样会显得更醒目，技术人员一看就知道数量级的大小。如果不是用符号表示，而是译成文字，则这些单位很容易被淹没在文字当中。而且，技术人员或者业内相关人员通常也更喜欢使用和看到用符号表示的单位。因此，译员应该形成这样的共识，就是尽可能把单位表示成符号。

例 6.8：

原文：5 HP TEFC motor rated for 460 VAC, 3 ph, 60 HZ operation.

译文：5hp 全封闭风扇冷却（TEFC）电机，额定工作条件 3 相、60Hz 交流 460V。

分析：原文里有较多的缩写，包括 HP、TEFC、VAC、ph 和 HZ，其中除了 TEFC 不常用，全拼为 Totally Enclosed Fan Cooled，意思为"全封闭风扇冷却"外，其他都是译员必须记住的词。

HP：horse power，马力，相当于 735W，其规范写法是 hp。

VAC：volt alternate current，交流电压。

ph：phase，相。

HZ：规范写法是 Hz，功率单位赫兹的英文，通常不译。

小结：

提及计量单位的表达，这里简单总结一下计量单位符号字母的大小写规则。

计量单位符号的书写规范

各种计量单位符号是用大写还是小写字母有严格的规定：

凡是来源于人名的单位符号第一个字母必须用大写字母（或只用一个大写字母），如：Hz（赫兹）、V（伏特）、A（安培）、Pa（帕斯卡）等。而其他非来源于人名的单位符号除一个特例外，一律用小写字母，如：m（米）、g（克）、s（秒）等，只有一个特例，就是表示"升"的符号，在容易与阿拉伯数字 1 混淆的场合必须用大写字母 L，如 5 升必须写成 5L，而在不容易与阿拉伯数字 1 混淆的场合，大小写均可，如 5 毫升可以写成 5ml。另外，有一个容易被误解的，就是发光强度的单位"坎德拉"不是人名，英文是 candle（本义为"烛光""蜡烛"），缩写为 cd，首字母不大写。

此外，表示倍率关系的字头也有严格规定，例如表示一千必须用小写的 k，而不应当用大写的 K，有的用错了还会引起歧义，例如 mW（毫瓦）与 MW（兆瓦）、mΩ（毫欧）与 MΩ（兆欧）、Nm（牛顿·米，力矩单位）与 nm（纳米，长度单位）等，导致差之毫厘，失之千里。

根据上述规定，kW（千瓦）写成 KW，kWh（千瓦时）写成 KWH 或 kwh，km（千米）写成 KM 或 kM，mm（毫米）写成 MM 或 mM 等都是不规范的。但这种现象在社会上比比皆是，屡见不鲜。

（续表）

> **常用电力计量单位的大小写规则**
> 在工作中我们常遇到很多人在写电力常用计量单位时，大小容易搞错，比如 kV 写成 KV 或干脆写反成 Kv，很不规范，下面将一些常用的计量单位字母的大小做个介绍。
> **电力常用的一些计量单位及其大小写：**
> （1）电压 kV、电阻 kΩ、电流 mA、μA、kA、有功功率 kW、无功功率 kvar、视在功率 kVA、电容 F、p、μF、频率 Hz 等。其中的 k、m、μ、var、p、z 均为小写字母，其余为大写字母。
> （2）MΩ、MW、Mvar、MVA、MPa 中的 M 为大写字母。
> （3）不少作者往往大小写不分，如将"mA（毫安）"写作"MA（兆安）"，将"MW（兆瓦）"写作"mW（毫瓦）"，则大错特错，贻笑大方。
> **外文字母的处理**
> 外文字母的大小写、正斜体、上下角标，必须书写清楚，区别明显。如千瓦 kW、千瓦时 kWh、千伏安 kVA、千乏 kvar、千安 kA、毫安 mA、微安 μA、千欧 kΩ、兆欧 MΩ、微法 μF、皮法 pF、兆帕 MPa、六氟化硫 SF_6 等。

例 6.9：（数量级的表达）

原文： 陶瓷蓄热体：陶瓷蓄热体采用 LANTEC 公司专利产品，该产品是用于 RTO 设备的最好蓄热产品。陶瓷蓄热体其特点是比表面积大（680M2/M3），阻力小，热容量大 0.22BTU/lb °F，耐温高可达 1200℃，耐酸度 99.5%，吸水率小于 0.5%，压碎力大于 4kgf/cm3，热胀冷缩系数小，为 4.7×10^{-8}/℃，抗裂性能好，寿命长。

原译： Ceramic heat accumulator: it shall adopt the patented product of LANTEC, which is the best heat accumulator for RTO equipment. The ceramic heat accumulator is characterized by large specific surface area (680M2/M3), small resistance, large heat capacity (0.22BTU/lb °F), high temperature resistance (up to 1200 ℃), acid resistance (99.5%), low absorption rate (0.5%), large crushing force (more than 4kgf/cm3), small thermal expansion coefficient (4.7×10-8/℃), good anti-cracking performance and long service life.

改译： Ceramic heat accumulator: it shall adopt the patented product of LANTEC, which is the best heat accumulator for RTO equipment. The ceramic heat accumulator is characterized by large specific surface area (680m²/m³), small resistance, large heat capacity (0.22BTU/lb °F), high temperature resistance (up to 1200 °C), acid resistance (99.5%), low absorption rate (0.5%), large crushing force (more than 4kgf/cm³), small thermal expansion coefficient (4.7 × 10-8/°C), good anti-cracking performance and long service life.

分析：本例的译文总体上问题不大，但是有些小细节必须注意。这些细节就是原文数字和单位的表达，如 M2/M3、kgf/cm3 和 4.7 × 10-8/℃，对数字和单位比较敏感的译员，尤其是有理工背景的译员，应该知道 M2/M3 就是 m^2/m^3，kgf/cm3 就是 kgf/cm^3，4.7 × 10-8 就是 4.7×10^{-8}，这些细节都是体现译员基本功的地方，即使原文表达有误，也建议译员在译文中把错误纠正过来，这样的处理法，在客户处具有印象加分作用。

例 6.10：（计量单位的中文写法）

原文：75 寸用户反映两半模硫化轮胎模口出边跟模口出台，经过在整副模具上压铅 220T 后靠模具外螺栓附近的数据如下：

译文：Users of 75″ vulcanizing machine reflected that two-piece mould vulcanized tyres have an edge and a terrace, and after 200T lead pressing on the whole mold, the data for outer bolts near the mold are as follows:

分析：以上的 75 寸，不能译成 75 cun。

这里的 75 寸，经询问客户，确定是指 75 英寸。企业内经常会把英寸称为寸，大家将"寸"理解成"英寸"，并用""""表达，相当于 inch（缩写为 in）。最常见的例子就是手机、电脑、电视屏幕的几寸，都是对角线的英寸数。如果出现尺，则是英尺的意思，用"'"表达，相当于 feet/foot（缩写为 ft）。

另外：

呎 = 英尺，中文拼音：yīng chǐ

吋 = 英吋，中文拼音：yīng cùn

哩 = 英里，中文拼音：yīng lǐ

浬 = 海里，中文拼音：hǎi lǐ

以上几个词，在网络里查到的拼音都是单音，但在辞海里是双音，也就是一个汉字发两音，这是非常罕见的现象。

（四）数字和字母表达特例

例 6.11：

原文：另外有 4 个 V4020-DZG 3/4' 滑阀及 1 个 V4020-DZG 1/2' 滑阀给了 3.5 公斤的风压（也拿了 7 公斤的风做过实验），阀芯伸缩极为困难，基本不动，导致阀门所对应的动作极为缓慢。

译文：Moreover, four V4020-DZG 3/4" slide valves and one V4020-DZG 1/2" slide valve have an air pressure of 3.5kg/cm^2 (7kg/cm^2 air pressure is also tested), valve element extends or retracts difficultly, can hardly moves, so the valves move very slowly.

分析：本例注意点：

（1）3/4'=3/4' = 四分之三英寸；1/2'=1/2' = 二分之一英寸。

（2）3.5 公斤的风压 = air pressure of 3.5kg/cm^2。

本例的以上数量，不仅要理解准确，还要注意上下标的表达，如 1/2' 和 kg/cm^2。

例 6.12：

原文：The consumption of hydrogen ions that occurs as iron dissolves leaves a preponderance of hydroxide (OH-) ions in the water. The ferrous ions react with them to form green rust:

Fe2+ + 2OH- → Fe(OH)2

That isn't the end of the story. The ferrous ions also combine with hydrogen and oxygen in the water to produce ferric ions:

4Fe2+ + 4H+ + O2 → 4Fe3+ + 2H2O

译文：当铁融解时，氢离子的消耗使得水中的氢氧根（OH-）离子占优势。亚铁离子与氢氧根反应，就生成绿色的铁锈：

$Fe^2+ + 2OH- \rightarrow Fe(OH)_2$

但这还没完。亚铁离子也会也与水中的氢和氧结合，生成铁离子：

$4Fe^2+ + 4H+ + O_2 \rightarrow 4Fe^3+ + 2H_2O$

分析：本例具有一定的专业性。翻译过程中，不仅要注意句子的理解和翻译，还要注意方程式中化学元素字母的大小写，以及化学元素后面数字和正负号的上下标格式。当然，如果译员不了解这些字母和数字的书写规则，也只能直接照抄。

例 6.13：

原文：Irradiance for Ukraine between 1.150-1.550 kWh/m²/a (Kiev: approx. 1.400 kWh/m²/a)

原译：乌克兰光照值范围：1.150-1.550 kWh/m²/年（基辅：约 1.400 kWh/m²/年）

改译：乌克兰光照值范围：1,150-1,550 kWh/m²/年（基辅：约 1,400 kWh/m²/年）

分析：本句的背景是太阳能发电，句中三个数字（1.150、1.150、1.400）都有三位小数，乍一看似乎都没什么问题。但如果细心一点，会不会发现这样的数值有点太小，每年每平方米的光照值仅有不到 2kWh，也就是一年一平方米光照值不到两度电，这样的太阳能发电也太没效率了吧。带着这个疑问，到网络上查"各地日照强度值""全球日照强度值排名""日照强度大小""日照强度单位"等，好像都查不到具体某地一年的日照强度的具体数据。

后来，考虑到这里的日照强度单位为 kWh，所以在百度里查"热带日照强度 kWh"，结果发现有一条链接里出现"某地的日照强度是 1784kWh/m²/year"的内容，而且这个链接里还有这样的回复："峰值日照时长是指在标准条件下的日照小时数，标准条件下的光照强度是 1000W/m²，因此这个换算是正确的。"而且，通过其他渠道，也可得知太阳辐射强度值通常都是四位数比较多，因此，可以确定本句里的 1.150、1.150、1.400 其实分别是 1,150、1,150、1,400，也就是说，这里也把逗号写成了点号。其实，以上数字表达法，正是本节开头的"数字量级表达法"里说明的一种数字表达法。

（五）金额表达特例

例 6.14：

原文：

百	十	万	千	百	十	元	角	分	注
X	X	1	2	3	4	5	0	0	
小计：12,345.00									

译文：

Amount	Note
RMB Twelve Thousand Three Hundred and Forty Five Only	
Subtotal: RMB12,345.00	

分析：本例原文的金额写法，是中国特有的方式，因为中国的数字单位，如一、二、三……十、百、千、万、亿都是一个字，很整齐。但你见过欧美国家这样写金额吗？欧美国家的数字都是一个单词，少则两三个字母，多则七八个字母，很不整齐而且很宽，如果也用这种方式，那麻烦可大了。

所以，这种中文表格式的金额写法，在翻译成英文时，不能再生搬硬套地使用表格，而是要把所有小格子合成一格。在本例中，就应该像以上译文一样进行处理。

四、常见标准、指令、条例、协会的缩写

在科技类翻译中，通常会碰到一些标准的代号，很多代号后面还会跟一串数字，如：

GB/T 15753　　圆弧圆柱齿轮精度

ISO18001　　职业安全卫生管理体系

这里将国际上一些主要的标准、指令、条例、行业认证机构或行业标准协会的缩写列示如下，这些缩写的意思，大家尽可能记住，如果记不住，也要大概知道是什么方面的意思，以节约时间。

缩写	全称	中文参考名称
ANSI	American National Standards Institute	美国国家标准协会
ASTM	American Society for Testing and Materials	美国材料与试验协会
BS	British Standards (Institution)	英国标准（化协会）
CCC	China Compulsory Certification	中国强制认证

（续表）

缩写	全称	中文参考名称
DIN	Deutsches Institut für Normung (German Standardization Institute)	德国标准化协会
GB	Guo Biao	国标（中国国家标准）
IEC	International Electrotechnical Commission	国际电工技术委员会
IEEE	Institute of Electrical and Electronic Engineers	（美国）电气与电子工程师协会
ISO	International Standardization Organization	国际标准化组织标准
JIS	Japanese Industrial Standards	日本标准
MSDS	Material Safety Data Sheet	物质安全资料表（材料安全数据表）
NEN	Netherlands Standardization Institute	荷兰标准化协会
OHSAS	Occupational Health and Safety Assessment System	职业健康安全评估体系
REACH	REGULATION concerning the Registration, Evaluation, Authorization and Restriction of Chemicals	（欧盟）化学品注册、评估、授权和限制条例
RoHS	Restriction of the use of certain Hazardous Substances	（欧盟）限制在电子电气产品中使用有害物质的指令
SGS	Societe Generale de Surveillance	瑞士通用公证行
UL	Underwriters Laboratories	美国保险实验室
WEEE	Waste Electrical and Electronic Equipment (WEEE) Directives	（欧盟）关于报废电子电气设备的指令

本节课后练习

翻译下列句子：

1. The Johannesburg Stock Exchange (JSE) is the largest stock market in Africa and the 14th largest stock market in world. At the end of 2012, the JSE had a market capitalisation of approximately of US$900 billion (R12,510 billion).
2. 康佳55寸液晶电视功耗为100至150W，待机功率约0.5W。

3. 一台 10 吨蒸汽锅炉，要求蒸汽供给压力 1MPa，锅炉热效率 75%，所用煤低发热量 5500 大卡，一吨煤能产多少汽？

4. 碳酸钠分子式为 Na_2CO_3，常温下为白色无气味的粉末或颗粒，有吸水性，露置空气中逐渐吸收 1mol/L 水分（约 15%）。其水合物有 $Na_2CO_3·H_2O$、$Na_2CO_3·7H_2O$ 和 $Na_2CO_3·10H_2O$。

第七章　各行业语篇翻译分析

本部分内容为科技类文件的实战分析。在学习本部分内容时，建议大家先只看原文，不看译文，尝试着自己将原文译出，或者在头脑中构思一下该怎么译，然后再看参考译文。最后，再看页尾对相关知识点或难点的翻译解释或背景知识。这样可以让大家提高学习效果。

第一节　通用类

一、中译英

2016 年**国内**糊树脂市场**走势震荡**，整体价格变化幅度**较大**。① 年末价格较年初上涨 3000 元/吨左右，涨幅接近 50%。② 整体走势明显分为四个阶段：

第一阶段（1—2 月）：年初开局，国内糊树脂市场走势平稳，价格几无变化。但由于整体**行情**低迷，**厂家**亏损严重，下游多处于春节前后的低负荷开工，**市场**

---

① 2016 年**国内**糊树脂市场**走势震荡**，整体价格变化幅度较大 = 2016 年中国糊树脂市场不稳定/反复无常，整体价格变化剧烈：Chinese paste resin market is volatile in 2016, with a severe change of overall price.
　　分析：本句的"国内"不宜字面译成 domestic，而是要理解其实际意思（"中国［国内的］"），译成 Chinese，否则会引起理解障碍；后面的"国内"也采用类似译法。另外，"走势震荡"也不宜字面翻译，宜译成 is volatile。后面的"整体价格变化幅度较大"则宜译成独立主格结构。

② 年末价格较年初上涨 3000 元/吨左右，涨幅接近 50%。　The yearly ending price is 3000 CNY / ton higher than the yearly beginning price, up by nearly 50%.
　　年末/年初：yearly ending/ yearly beginning。
　　说明：yearly beginning 里的 yearly 也可省略，以避免重复。
　　涨幅/跌幅……：up/ down by ...

交投氛围僵持。① 在此阶段，国内主流厂家维持较为稳定的开工负荷②，供过于求的市场态势③难以逆转。

第二阶段（3—5月）：随着节日气氛逐渐散去④，下游厂家开工不断提升，采购气氛稍显活跃，恰逢此时，国内主流厂家——天津渤天化工确定停车搬迁，整体市场供应量萎缩⑤。两相作用下，市场出现短暂的货源紧张局面⑥。下游正处于恢复开工、交付新单的阶段，采购较为积极，推动行情一路上行。⑦ 与此同时，国内电石价格出现年内第一次上扬，对糊树脂的成本支撑力度增强⑧。

第三阶段（6—8月）：随着价格的不断上行，国内糊树脂装置⑨开工小幅提升，

---

① 但由于整体行情低迷，厂家亏损严重，下游多处于春节前后的低负荷开工，市场交投氛围僵持。However, the overall downturn market brings severe loss to manufacturers, most downstream industries operate with low load around Chinese New Year and the trading atmosphere is deadlocked.

  分析：本句中各小句均不宜字面翻译，而是要经过一定的意思转换后再译出，其转换如下：
    由于整体行情低迷，厂家亏损严重 = 整体下行的市场给厂家（= 制造商）带来严重亏损；
    下游多处于春节前后的低负荷开工 = 下游行业在中国新年前后以低负荷开工；
    市场交投氛围僵持 = 交易氛围僵持。
    低迷：downturn。 亏损严重：severe loss。 厂家：manufacturer。
    市场交投氛围：trading atmosphere。 僵持：be deadlocked。

② 维持较为稳定的开工负荷 = 维持稳定负荷的生产：maintain production in a stabilized load。

③ 供过于求的市场态势 = 过量供给的态势：status of over-supply。

④ 随着节日气氛逐渐散去：As the festival atmosphere dissipates gradually。
  节日气氛：festival atmosphere。
  散去 = 褪去；消散：fade；dissipate。

⑤ 恰逢此时，国内主流厂家——天津渤天化工确定停车搬迁，整体市场供应量萎缩。Meanwhile, one main-stream manufacturer— Tianjin Botian Chemical decides to shut down and relocate, dwindling the overall supply accordingly.

  分析：这里的"停车"意思为"停产"（shut down; stop production），"停车"还有另一意思"停机"，其相应译文为 stop。另外，"整体市场供应量萎缩"这一小句是主句的补充说明，可译成非谓语动词结构 dwindling ...

⑥ 货源紧张局面 = 供货紧张：tight supply。

⑦ 下游正处于恢复开工、交付新单的阶段，采购较为积极，推动行情一路上行 = 下游企业正在恢复开工、交付新单，采购氛围变热，推动市场一路上行 Down-stream enterprises are restoring the production and delivering new orders, so the purchasing air gets hot, driving a market rise all the way.

⑧ 对……支撑力度加强 = 加强……的基础：underpin。

⑨ 糊树脂装置：paste resin plants。其中，装置 = 工厂：plants。

市场供需关系悄然发生逆转。但受到成本的支撑，价格降幅有限。① 供需双方进入僵持博弈阶段②。但后期市场的走势可以说超出了双方的预期。

第四阶段（9—12月）：在这一时期，国内糊树脂市场可谓风生水起，价格一路飙升③，连续突破近期的高位④。分析原因：**一方面**，由于西北环保检查力度突然增强，当地电石生产装置**开工**明显下调，货源供应紧张，特别是作为糊树脂主产区的**华北、东北**等地，到货价格连续上扬，成本支撑力充足⑤；另一方面，9月21日实行的公路运输新规，对超载超限的检查力度空前⑥，无论是原料采购还是产品销售的成本大幅度上涨，进一步**增强**⑦了糊树脂市场的支撑力度；第三

---

① 但受到成本的支撑，价格降幅有限 = 但成本因素限制了价格下降幅度：But the cost factors limit the dropping of prices。

② 进入僵持博弈阶段 = 进入对峙游戏：enter a stand-off game。

③ 一路飙升 = 稳定升高：soar steadily。

④ 高位：high (n.)。近期高位：recent highs。

说明：这里将 high 当名词使用，是典型的熟词僻义。

⑤ 一方面，由于西北环保检查力度突然增强，当地电石生产装置开工明显下调，货源供应紧张，特别是作为糊树脂主产区的华北、东北等地，到货价格连续上扬，成本支撑力充足 Firstly, the inspection for environmental protection in Northwest China is intensified suddenly, the operation rate of local calcium carbide plants dropped obviously, and the supply is tight. Therefore, in major production regions of paste resins, especially in North and Northeast China, the delivered prices rise continuously, strongly strutting the cost.

分析：这一句是对前一句的原因说明，其原因不是只有两点，而是有三点，所以，尽管原文表达成"一方面……另一方面……第三方面……"，但还是宜将"一方面……另一方面……"译成 Firstly... Secondly ... 以与后面的 Thirdly 相互呼应。

另外，"特别是作为……"开始的后半句，是前半句的结果，而且原文句子较长，所以，可以将前半句和后半句单独成句，但后半句之前宜加个 Therefore 等表示结果的词。

开工 = 开工率：operation rate。

华北、东北：North and Northeast China。说明：出于避免重复的原则，这里可省略了前一个 China。

……成本支撑力充足 = ……强力支撑着成本：...strongly strutting the cost（非谓语动词结构表伴随状态）。

⑥ 9月21日实行的公路运输新规，对**超载超限**的检查力度空前 the new regulations on road transportation implemented from Sep. 21 pose an unprecedented pressure on the overloading and overruns of vehicles.

分析：本句可理解为"9月21日实行的公路运输新规对超载超限车辆施加空前的压力"，其中"超载超限"的意思是超过（额定）负荷和超过车辆的载货尺寸（长、宽、高）和载货质量超过道路等的规章制度规定的限度，所以理解成超载和超过限度的汽车，译成 overloading and overruns of vehicles。

⑦ 增强：reinforce。说明：出于避免重复的原则，这里不再使用 intensify，而是换用 reinforce。

方面,<u>国内供给侧改革初见成效,从基础的煤炭开始,国内化工市场行情在四季度出现明显的回暖</u>①,受益于此,不管是从上游成本方面还是下游信心方面,均跟糊树脂市场形成良**好的互动**②。

Chinese paste resin market is volatile in 2016, with a severe change of overall price. The yearly ending price is 3000 CNY/ton higher than the yearly beginning price, up by nearly 50%. The overall trend can be obviously divided into four stages:

Stage I (Jan – Feb): Chinese paste resin market is stable in the early year, with little price change. However, the overall downturn market brings severe loss to manufacturers, most downstream industries operate with low load around Chinese New Year and the trading atmosphere is deadlocked. During this stage, main-stream manufacturers in China maintain production in a stabilized load and the status of over-supply cannot reverse.

Stage II (Mar – May): As the festival atmosphere dissipates gradually, the working load of down-stream manufacturers increases continuously, and the purchasing will gets slightly active. Meanwhile, one main-stream manufacturer — Tianjin Botian Chemical decides to shut down and relocate, dwindling the overall supply accordingly. These two factors result in a tight supply temporarily in the market. Down-stream enterprises

---

① 国内**供给侧改革初见成效**,**从**基础的煤炭**开始**,国内化工市场行情在四季度出现明显的**回暖**
Thirdly, the supply-side reform in China witnesses its preliminary result, the chemical market in China, following the coal market, rebounds notably since last Q4.

  分析:本句译文意思相当于"国内供给侧改革见到了成效,国内化工市场在煤炭市场之后,也在去年第四季度出现明显的反弹",其中"基础的"为意思冗余部分,可以不译。

  供给侧改革:supply-side reform。初见成效:witness its preliminary result。

  从……开始 = 在……之后:following。回暖 = 反弹:rebound。

  四季度:the fourth quarter = Q4。类似的表达还有:W2(第2周)、Day3(第3天)。

② **受益**于此,不管是从上游成本方面还是下游信心方面,均跟糊树脂市场形成**良好的互动**。
Benefiting from these factors, a benignant circle forms between both the upstream cost and the downstream confidence and the market of paste resins.

  分析:本句的意思是成本方面和信心方面都跟糊树脂市场形成了良性循环,所以这里要译成 a benignant circle forms between ... and ... and the market of paste resins。

  受益于……:benefiting from ...

  良好的互动 = 良性循环:benignant circle。

are restoring the production and delivering new orders, so the purchasing air gets hot, driving a market rise all the way. Meanwhile, prices of calcium carbide in China rise for the first time this year and underpin the cost of paste resins.

Stage III (Jun – Aug): with the continuous increase of prices, the operation rate of paste resin plants in China goes up slightly and the supply and demand situation reverses quietly. But the cost factors limit the dropping of prices. The suppliers and buyers enter a stand-off game. However, it can be said that the subsequent market trend goes beyond the expectations of both sides.

Stage IV (Sep – Dec): the market of paste resins in China springs up this stage and the prices soar steadily and break the recent highs again and again. The reasons include: Firstly, the inspection for environmental protection in Northwest China is intensified suddenly, the operation rate of local calcium carbide plants dropped obviously, and the supply is tight. Therefore, in major production regions of paste resins, especially in North and Northeast China, the delivered prices rise continuously, strongly strutting the cost. Secondly, the new regulations on road transportation implemented from Sep. 21 pose an unprecedented pressure on the overloading and overruns of vehicles. The purchasing cost of raw materials and the sales cost of products both rise sharply, which further reinforce the support to paste resin market. Thirdly, the supply-side reform in China witnesses its preliminary result, the chemical market in China, following the coal market, rebounds notably since last Q4. Benefiting from these factors, a benignant circle forms between both the upstream cost and the downstream confidence and the market of paste resins.

二、英译中

The results for the three main regions[①] are reported in Figure 5.9. Exports of

---

① three main regions：三个主要地区。这里指的是欧洲、东亚和北美三个地区。

KETs-related① products by the EU-28② face very different competition on international markets. In advanced manufacturing technologies for other KETs③, most EU-28 exports (64%) concern products for which trade is **characterised by** quality competition④. For almost all these products, the EU-28 has a quality advantage; in other words, it is able to gain a **positive trade balance**⑤ based on superior product quality. In nanotechnology, industrial biote-chnology and advanced materials, only 23 % to 34 % of EU-28 exports are based on⑥ quality competition. Although the majority of EU-28 exports in these KETs is characterised by price competition, most of these exports **benefit from**⑦ price advantages. This means that Member States specialise in those

---

① KETs-related = key enabling technologies-related：（与）关键使能技术相关的。（KET 在上文出现过全称）

说明：英译中时，除了一些在中文里被普遍接受的英文缩写（如：DNA、CPU 等）外，其他英文缩写一般都要译成中文。

② EU-28：欧盟 28 国。

③ other KETs：其他关键使能技术（关键使能技术里的一个分类）。

④ most EU-28 exports (64 %) concern products for which trade is characterised by quality competition 欧盟 28 国的大多数出口产品（64%）面临的是质量竞争

分析：本句主干为 most EU-28 exports concern products，后面 for which 引导的是 products 的定语从句。本句谓语 concern 本义为"关系到""与……相关"，这里可译为"是"。本句可字面翻译为"欧盟 28 国的大多数出口产品（64%），其产品贸易都是以质量竞争为特征"，但这样的译文不如参考译文简洁顺畅。参考译文相当于将 be characterised by 译成"面临"。

⑤ positive trade balance：正贸易差额；贸易顺差。

⑥ are based on：基于。这里理解为"面临"。本文中的 be characterised by 和 focus on 也是类似的意思。

⑦ benefit from：受益于；得益于；因……而得到好处。这里理解为"具有"。

price-sensitive products for which a **cost-efficient production** in the EU is possible.① In **photonics** and micro-/ nanoelectronics②, most of the products exported by the EU-28 are in price competition (89 % and 94% respectively), and for the majority of these products the EU has no price advantage.

North America reports a strong focus on exports which face quality competition: **it** relies on a quality advantage in international trade in the fields of photonics (78 % of all exports in this KET) and nanotechnology (54 %).③ In micro- and nano- electronics, 41 % of North America's exports fall into **this category**④, while 15 % are characterised by quality competition, without having a quality advantage. In the other three KETs, exports from North America mainly face price competition, with a price advantage over their main competitors.

East Asia's trade in KET-related products is strongly focused on price competition. In five KETs —— industrial biotechnology, nanotechnology, micro-and nanoelectronics,

---

① This means that Member States **specialise in** those **price-sensitive products** for which a **cost-efficient production** in the EU is possible.

原译：这意味着，成员国专注于能在欧盟以成本有效方式生产的价格敏感型产品是可行的。

改译：这意味着，成员国专注于可在欧盟内实现具有成本效益的生产的价格敏感型产品。

分析：本句主干为 This means that Member States specialise in those price-sensitive products，其中 that Member States specialise in those price-sensitive products 是 that 引导的宾语从句，for which a cost-efficient production in the EU is possible 则是 those price-sensitive products 的定语从句。原译将主句的宾语从句理解为 Member States specialise in those price-sensitive products is possible，句子结构理解出现错误。

specialise in：专攻；专注于；专门研究。

price-sensitive product：价格敏感型产品。

cost-efficient production：具有成本效益的生产。

② photonics and micro-/nanoelectronics：光子和微电子/纳米电子领域。photonics：*n.* 光子；光子学。

③ North America reports a strong focus on exports which face quality competition: **it** relies on a quality advantage in international trade in the fields of photonics (78 % of all exports in this KET) and nanotechnology (54 %) 北美出口产品面临的主要是质量竞争：其中光子这一关键使能技术领域总出口的 78%，以及纳米技术这一关键使能技术领域的 54% 的国际贸易，都有赖于其质量优势。

分析：本句主干为 North America reports a strong focus on exports，后面的 which face quality competition 是 exports 的定语从句，冒号后为主句的补充说明。另外，第二个括号里，54% 后面省略了与第一个括号相同的成分，即 of all exports in this KET。其中，it 指代的是 North America。

④ this category：这一类别。这里指的是上一句所讲的有赖于质量优势的关键使能技术领域。

advanced materials and advanced manufacturing technologies for other KETs — East Asia benefits from a price advantage, in other words a cost-efficient production. Photonics is the only area where East Asia's exports are under major pressure①, as most of its products face price competition but cannot compete on a price advantage. In each KET, the share of KET-related products exported from East Asia which are in markets dominated by quality competition is lower than from North America, ranging from 10 % (micro-and nanoelectronics) to 29 % (photonics).② The majority of these exports do not have a quality advantage.

When examining the development of competition types **by** KET over time, no clear trends for the EU-28 emerge.③ In advanced manufacturing technologies for other KETs, the share of EU exports based on quality competition and quality advantage increased during the 2000s, while the share of exports④ based on price competition and price advantage decreased. In photonics, the share of EU exports in markets with price competition which could profit from an EU price advantage has fallen **substantially** in the last ten years, while the share of exports facing price competition without a price

---

① major pressure：巨大压力。major：*adj.* 重大的；重要的；主要的。

② In each KET, the share of KET-related products exported from East Asia which are in markets dominated by quality competition is lower than from North America, ranging from 10% (micro-and nanoelectronics) to 29 % (photonics). 在各个关键使能技术领域中，东亚出口的、主要面临质量竞争的关键使能技术相关产品的比例介于10%（微电子和纳米电子）和29%（光子），低于北美。

分析：本句主干为 the share of KET-related products exported from East Asia is lower than from North America，其中 than 后省略了相同成分 the share of KET-related products。后面的 ranging from ... to ... 是非谓语动词结构，是对 share 的补充说明。其中 in markets 是 dominated 的地点状语，但可以省译。

③ When examining the development of competition types **by** KET over time, no clear trends for the EU-28 emerge. 在考察按关键使能技术分类的竞争类型随时间变化情况时，未发现欧盟28国呈现出明显的趋势。

分析：本句逗号后的内容是主句，前面 when 引导的是时间状语从句。其中 development of A by B over C 的意思是"按B分类的A随C的变化情况"，by 的意思是"按……（逐个）分类"的意思。

④ exports = European exports。

advantage has increased.①

  三个主要地区的竞争类型计算结果如图 5.9 所示。欧盟 28 国关键使能技术相关产品出口在国际市场上面临的竞争差异较大。在其他关键使能技术先进制造技术领域，欧盟 28 国的大多数出口产品（64%）面临的是质量竞争。欧盟 28 国在几乎所有这些产品上都具有质量优势，换句话说，就是能够凭借更高的产品质量获得贸易顺差。在纳米技术、工业生物技术和高级材料领域，欧盟 28 国在这三个关键使能技术领域的大部分出口以价格竞争为主，但这些出口产品大多数都具有价格优势。这意味着，成员国专注于可在欧盟内实现具有成本效益的生产的价格敏感型产品。在光子和微电子/纳米电子领域，欧盟 28 国出口的大部分产品面临的是价格竞争（分别为 89% 和 94%），但大多数产品均无价格优势。

  北美出口产品面临的主要是质量竞争：其在光子这一关键使能技术领域总出口的 78%，以及纳米技术这一关键使能技术领域的 54% 的国际贸易，都有赖于其质量优势。在微电子和纳米电子领域，北美总出口的 41% 都属于这种类别，其中 15% 面临的是无质量优势的质量竞争。在其他三个关键使能技术领域，北美出口面临的，是与主要竞争对手相比有价格优势的价格竞争。

  东亚的关键使能技术相关产品贸易面临的主要是价格竞争。东亚在工业生物技术、纳米技术、微电子和纳米电子、高级材料和其他关键使能技术先进制造技术这五个关键使能技术领域的贸易具有价格优势，换句话说，其生产具有成本效益。东亚出口唯一面临巨大压力的是光子领域，因为其大多数光子产品面临着价格竞争，但又不具备价格优势。在各个关键使能技术领域中，东亚出口的、主要面临质量竞争的关键使能技术相关产品的比例介于 10%（微电子和纳米电子）和 29%（光子），低于北美。大多数这些出口产品不具备质量优势。

---

  ① In photonics, the share of EU exports in markets with price competition which could profit from an EU price advantage has fallen **substantially** in the last ten years, while the share of exports facing price competition without a price advantage has increased. 在光子领域，过去十年里，面临价格竞争且能借助欧盟价格优势盈利的欧盟出口产品的比例大幅下降，而面临价格竞争但无价格优势的出口产品的比例则有所增加。

  分析：本句的主干为 the share of EU exports in ... which ... has fallen substantially in ... while the share of exports facing ... without ... has increased，其中第一个 EU exports 后面的前一个 in ... 和 which 从句都是 EU exports 的定语，而 in the last ten years 则为句子的时间状语，第一个 EU exports 后面的 facing ... 和 without ... 都是 EU exports 的定语。组织译文时，宜将 EU exports 作为一个整体译出，句子才会显得更紧凑。

在考察按关键使能技术分类的竞争类型随时间变化情况时，未发现欧盟 28 国呈现出明显的趋势。对于其他关键使能技术的先进制造技术领域，面临质量竞争且具备竞争优势的欧盟出口产品的比例在 2000 年有所增加，而面临价格竞争且具有价格优势的欧盟出口产品的比例则有所下降。在光子领域，过去十年里，面临价格竞争且能借助欧盟价格优势盈利的欧盟出口产品的比例大幅下降，而面临价格竞争但无价格优势的出口产品的比例则有所增加。

第二节　管理体系

一、中译英

3.8　责任部门对发生的不符合和潜在不符合情况①及时进行原因②分析，制定相应的纠正和预防措施③，并在规定时间内完成整改④。⑤

---

① 发生的不符合和潜在不符合情况：occurred and potential nonconformance。
发生的：*adj.* occurred。
区别：occur（发生；出现）；incur（招致；导致）。
不符合：nonconformance。
区别：nonconformance（不符合）；nonconformity（不合格）。不符合项（也称为不符合）是指产品或服务有某一项目达不到要求，而不合格则是指产品或服务整体质量达不到要求。一般而言，不合格比不符合更严重，不合格产品或服务中可能有一个或多个不符合（项），而有不符合项的产品不一定不合格，有时可让步接受。
② 原因：cause。
区别：cause：造成故障等某种不良后果的原因。reason：用于解释发生某种事情的原因。
③ 纠正和预防措施：corrective and preventive actions。
纠正措施：corrective action。纠正：correct / correction。
预防措施：preventive action。预防：prevent / prevention。
措施：action。区别：action 常实际采取的行动；measure 常指理论上的方案。
④ 整改：rectify / rectification。注意：完成整改 = 整改。
⑤ 关于情态动词：管理体系文件是企业或其他组织机构对其日常生产或运营中的各道程序的规范性要求，类似于国家和地方政府的法律法规，因此，其译法类似法律类文件，但正式程度较法律稍低。所以，管理体系文件也经常使用法律英文中常用的情态动词 may、can、will、shall、must 等。其中，有些原文里未使用"应""应当""宜"等词，实际也可根据上下文语境或逻辑推理，认为是强制性或推荐性规定，并使用情态动词 shall（应 / 应当）或 should（宜）。

3.9 <u>行政管理部</u>①负责环境<u>事故</u>、<u>事件</u>②的统计、报告、调查和<u>处理</u>③;并做好<u>事故统计、处理工作</u>④。

4. 工作<u>程序</u>⑤

公司<u>质量管理部</u>⑥应结合本单位的产品特点<u>制定</u>⑦《<u>不合格品管理制度</u>⑧》,对不合格品的<u>识别、分类、记录、隔离、评审和处置</u>⑨等环节做出具体规定。不

---

① 行政管理部:Administrative Department。其中 Department 通常可缩写为 Dept。

② 事故:incident/ accident。事件:event。

区别:incident:一般事故或事件。accident:经常指导致人物伤亡和损害的负面事故。event:重大事件。

③ 处理:treat / treatment。

区别:treat:应对、解决问题点。handle:用一定技巧应对和解决,常有正面赞许含意。

④ 做好事故统计、处理工作:(省译)。

分析:"并做好事故统计、处理工作"里的"统计"和"处理"在前文都已经说过一次,属于明显的啰唆重复,因此直接略去不译。

⑤ 程序:procedure。

区别:在管理体系中,以下两个词的译法相对固定:程序:procedure;工艺 / 过程:process。

⑥ 质量管理部 = 品管部:Quality Control Department(缩写 QC)。

说明:工厂里常见部门名称全称与缩写

质量保证部 = 品保部　　Quality Assurance Department = QA

质量管理部 = 品管部　　Quality Control Department = QC

研发部　　　　　　　　Research & Development Department = R&D

人力资源部 = 人资部　　Human Resource Department = HR

⑦ 制定:prepare/preparation;formulate/formulation;enact/enaction。

⑧ 不合格品:nonconforming product。管理制度:management system。

区别:nonqualified:(人)不称职的,无资质的,不合格的。

　　　nonconforming:(产品 / 服务)不合格,不符合要求。

⑨ 识别 identify / identification(辨认出某事物或现象)。

标识 label(为便于识别而做标记)。

分类 classify / classification

记录 record / recording

隔离 isolate / isolation

区别:separate:分开,分别。isolate:隔离,隔开。segregate:(人、生物)隔离 / 分离,种族隔离。

评审 review / reviewing

区别:review:评审。audit:审核。evaluate:评估。

处置 dispose of / disposal

区别:处理:treat—应对、解决问题点。

处置:dispose—妥当地对麻烦事做一定安排,如对垃圾、不良品、废弃物做出相应安排。

合格品的标识按《标识和可追溯性控制程序文件[①]》执行。

4.1 不合格品的分类

（a）严重不合格：经检验判定的批量不合格[②]，或造成较大经济损失、直接影响产品质量、主要功能、主要性能技术指标[③]（超过企业注册标准、影响测定结果可能导致诊断错误）不符合要求。[④]

（b）一般、偶然发生的不合格：一般或次要指标（非重要质量特性[⑤]）不符合要求。

[①] 标识和可追溯性控制程序文件 Control Procedures for Labeling & Traceability（文件名实词首字母要大写）。

标识：label / labeling。可追溯性：traceability。

[②] 批量不合格：batch nonconformity。批 / 批量：batch。

[③] 指标：indicator。

[④] 句子组织说明：

（1）像（a）点这样概念在前，名词列举性内容在后的定义类句子（"严重不合格：……的不合格，或……不符合要求"），在译成英文时，一般把后面内容译成英文的名词形式。如果原文名词有较长定语，英译时一般也译成定语。

（2）定义类句子的括号里的补充说明，一般也要根据上下文判断括号内容在句子中所扮演的成分，并以合适方式译出。如以上（a）点括号里的内容，就应当作后置定语译出；而（b）点括号内容，则应当作同位语译出。

（3）（a）点原文有错，对于出错的原文，译文一般按其真正本义译出。本句的本义应该为：

严重不合格：经检验判定的批量不合格，或造成较大经济损失、直接影响产品质量、主要功能的不合格，或者主要性能技术指标不符合要求（超过企业注册标准、影响测定结果可能导致诊断错误）。

[⑤] 非重要质量特征：unimportant quality characteristics。非重要：unimportant / insignificant。

4.2 报告

（a）生产过程中发现不合格产品，按照《偏差处理管理制度》进行；①

（b）检验中发现不合格品，应按《检验操作规程》规定记录在相应检验报告中，并按《质量检验管理制度》进行处理，必要时查找原因并处理②。

4.3 进货不合格品可采用"限定性合格"方式使用③或退换货④。

4.3.1 质量管理部对不合格物料出具不合格报告单，供应部将不合格物料放置于不合格区，由采购中心⑤办理退货处理手续⑥。

4.3.2 物料作"限定性合格"使用时，由质管部做出有条件的"限定性合格"

---

① 生产过程中发现不合格产品，按照《偏差处理管理制度》进行：Nonconforming products found in the product process shall be treated in accordance with *Deviation Treatment Management System*.

语态分析：

（1）对于施动者不明确或无需说明的中文句子，通常以受动者作为句子主语，采用被动语态。或者说，以主动语态不好表达，转换成被动语态却比较好表达的句子，宜采用被动语态。如4.2的（a）和（b），以及4.3、4.3.1后最后一小句，和4.3.2的全部。

（2）对于施动者很明确或需加以说明的中文句子，通常以施动者为句子主语，采用主动语态。如3.8、3.9、4.0和4.1的（a）和（b）。

时态分析：

管理体系常用的时态为现在时、过去时以及少量将来时。现在时用于一般陈述句[如3.8、3.9和4.1的（a）和（b）]，包括假设性陈述句[如4.2的（a）和（b）]。

将来时用于表示将来可能出现的结果，在管理体系中偶尔会使用到。如："质量管理部、生产部审查供方回复的纠正措施，若可接受，即可结案"。

② ……并按《质量检验管理制度》进行处理，必要时查找原因并处理。… treated as per *Management System for Quality Inspection*, and the causes must be found and the problems solved if necessary.

分析：当多个动词共用同一主语时，后续动词不仅可省略主语，而且 and 也可适当少用，比如 treated 前就少用了一个 and。另外，英语省略共同成分的情况非常见，"查找原因并处理"的意思相当于"查找原因并处理问题"，其中 the problems 和 solved 之间省略了共同成分 shall be。

③ "限定性合格"方式：in "restrictive conformity" manner。

④ 退货：return。换货：replace/replacement。

⑤ 采购中心：Procurement Center。注意：采购部、采购中心里的"采购"译成 procurement，不译成 purchasing。

⑥ 由……办理退货处理手续：the return / replacement procedures shall be completed by ...

使用的决定①，必要时生产部或研发中心进一步验证②，总工程师批准。

4.4 不合格半成品和成品的识别和处理③，处理方式有返工、报废④：

4.4.1 对于一般不合格品，经生产和质量管理部验证后，可要求生产人员立即返工，并将检验情况记录在半成品检验记录和成品检验记录内；对于分装、组装过程中判断为可返工的，可由质量管理部批准予以返工。⑤

（a）返工由生产部执行⑥，返工产品必须重新接受检验，并填写⑦相应检验记录；

---

① 物料作"限定性合格"使用时，由质管部做出有条件的"限定性合格"使用的决定……The decisions for using materials in "restrictive conformity" manner shall be made by QC....

　　分析：本部分"物料作'限定性合格'使用时"与后面的"有条件的'限定性合格'使用"意思一致，有啰唆之嫌，因此省译。

② 验证：verify / verification。

③ 不合格半成品和成品的识别和处理：The identification and treatment of nonconforming semi-finished and finished products。

　　分析：这一小句相当于 4.4 的小标题，小标题通常可译成非完整句子形式。这里将其译成名词形式。

　　半成品：semi-finished product。成品：finished product。

④ 返工：rework。报废：scrap。

⑤ 对于一般不合格品，经生产和质量管理部验证后，可要求生产人员立即返工，并将检验情况记录在半成品检验记录和成品检验记录内；对于分装、组装过程中判断为可返工的，可由质量管理部批准予以返工。Common nonconforming products may be required to be reworked by production personnel immediately after being verified by PD and QC and made records for the inpsection on the record of semi-finished and finished products. The products that are judged as reworkable in subpackaging and assembling processes shall be reworked with the approval of the Quality Control Department.

　　分析：以"对于"开头引出句子主题，或者以其他方式以一个小句引出主题（相当于句首有"对于"），作为后面小句的说明对象，通常可以将该主题作为译文主语，如以上译文第一句下加下画线部分。这种情况的第一小句有时会是"对于……的"之类的"的"字句，如本句分号后一句。但是，如果把这种"的"字句所述内容译成主语，一般要把"的"字句所述对象当作译文主语的核心，"的"字句的其他部分则当作定语状语等成分，如以上译文第二句下加下画线部分。

　　检验：inspect / inspection。

　　区别：inspect、check、test、detect 等词在管理体系中意思相对固定如下：

　　　　inspect/ inspection：检验。check：检查；查核。

　　　　test：试验；测试。detect：检测。

⑥ 本句中"执行"意思相当于"完成"。

⑦ 填写：fill in；complete。本句将"填写相应检验记录"理解为"做好相关检验记录"，未译成 fill in 或 complete。

（b）报废产品由生产部放置于不合格品区，按《不合格品管理制度》<u>处理</u>①。

4.4.2 <u>判定为严重不合格的产品，应由生产总监审核、总经理批准并在相应记录上签字确认，并由质量管理部按要求进行处理，填写不合格品的处理记录。</u>②

4.4.3 <u>每次返工应编制相应的返工操作规程，经生产总监或总工程师审批确定返工的影响并形成文件</u>③。

4.5 返工评审

（a）工艺部、质量管理部和生产部应对产品实行返工可能性提出意见，并预计由于处置和采取纠正措施而影响生产计划的程度，同时应对原材料或产品的

---

① （报废产品等的）处理：treat / treatment。

区别：treat / treatment：将错误、问题点给纠正过来，比如处理某异常情况、治病、处理汗水等。
　　　dispose / disposal：将废弃、不要的物品给处置掉，如将一次性用品、废品等给扔掉、报废掉。

② 判定为严重不合格的产品，应由生产总监审核、总经理批准并在相应记录上签字确认，并由质量管理部按要求进行处理，填写不合格品的处理记录。Those products which are judged as seriously nonconforming shall be reviewed by the Production Director and approved by the General Manager, with signatures on relevant records for confirmation. The Quality Control Department shall treat nonconforming products as required and fill in the treatment records for them.

分析：本段第一个逗号前部分其实是后两个小句的主语。而且照理来说，翻译本段时，应该把前三个小句译成有两个谓语的一个英文句子。但是，由于最后一小句的主语是第三小句中的施动者"质量管理部"，所以还是把后两小句合并处理，并以"质量管理部"作为主语。

③ 每次返工应编制相应的返工操作规程，经生产总监或总工程师审批确定返工的影响并形成文件：Relevant rework operating procedures shall be prepared for each rework, and reviewed by the Production Director or Chief Engineer for determining the influence of rework, and documented.

分析：本句共有两小句，其中第一小句不知道施动对象是谁，因此以受动对象"相应的返工操作规程"的译文 Relevant rework operating procedures 为主语，由于"经生产总监或总工程师审批确定返工的影响并形成文件"这一小句不好译，这里宜将这一句的两个动作"审批"和"形成文件"与前一句的译文共用主语，理解为"（返工操作规程）由生产总监或总工程师审批，以确定返工的影响，并形成文件"，这也正是参考译文的译法。

适用程度及对使用方面的影响是否满足法律法规的要求提出建议。①

（b）生产总监或总工程师应组织相关技术人员详细阐明原材料、半成品或成品质量问题的性质②。

（c）销售部应提供市场动态③，并对可能造成的影响程度发表意见。

（d）采购中心部应及时对在购进原材料的可能性和价格方面提供信息，并负责与供方联系以采取相应措施。

（e）质量管理部根据汇总意见填写不合格品报告单，写明原因和处置意见，交生产总监或总工程师批准，并通知相关部门组织实施，如需返工由工艺部会同技术人员制定返工工艺，按《返工管理制度》处理④。

---

① 工艺部、质量管理部和生产部应对产品实行返工可能性提出意见，并预计由于处置和采取纠正措施而影响生产计划的程度，同时应对原材料或产品的适用程度及对使用方面的影响是否满足法律法规的要求提出建议。The Technology Department, Quality Control Department and Production Department shall give suggestions on the possibility of rework, and predict the influence on production schedule caused by the disposal and corrective actions, and meanwhile, give advices on whether the applicability of raw materials or products and whether their influence on usage meets the requirement of laws and regulations.

　　分析：本段一整段就是一个句子，其句子结构分析如下，其中主干为加框部分：

　　工艺部、质量管理部和生产部 应对产品实行返工可能性 提出意见 ，并预计 由于处置和采取纠正措施而影响生产计划的 程度 ， 同时应 对原材料或产品的适用程度及对使用方面的影响是否满足法律法规的要求 提出建议 。

　　句子的其他部分，都是主干的补充成分，其中第一小句中"对……"是"提出意见"的对象，第二小句"由于……而……"是"程度"的定语，而第三小句的"对……及对……是否……"则是"提出建议"的对象。

　　具体翻译时，可以把句子主干先译出来，再把其他补充成分一一添加在适当的位置。值得一提的是，本段三个小句的补充成分，都可以在主干译文后用介词 on 译出。

② 性质：nature.

　　区别：nature 指的是事物的本质（内在）属性，比如猴子的性质是灵长类动物。property 指的是事物表现出来的特征、性能等，比如醋有与碱中和的性质特征。

③ 市场动态 = 市场趋势：market trend.

④ 如需返工由工艺部会同技术人员制定返工工艺，按《返工管理制度》处理：The Technology Department will prepare a rework process together with technical personnel if rework is needed, which shall be based on the *Management System for Rework*.

　　分析：本句中"按《返工管理制度》处理"的主语其实是"返工工艺"，这里将其译成 which 引导的非限定性定语从句，其中 which 是指前面一整句（"由工艺部会同技术人员制定返工工艺"），其实就相当于"处理"。

3.8 The responsible department shall analyze the causes of the occurred and potential nonconformance, make corresponding corrective and preventive actions and rectify them on time/within specified time period.

3.9 Administrative Department shall take charge of the counting, reporting, investigation and treatment of environmental incidents and events.

4. Working Procedures

Quality Control Department shall prepare a *Nonconforming Product Management System* in combination with the features of internal products, in which provisions for the identification, classification, recording, isolation, reviewing and disposal of non-conforming products shall be made specifically. The labeling of nonconforming products shall follow *Control Procedures for Labeling & Traceability*.

4.1 Classification of Nonconforming Products

(a) Serious nonconformity: Batch noncon-formity judged through inspection or nonconformity that causes great economic loss, directly affects the product quality or main functions, or nonconformance of major performance technical indicators (exceeding the registered standards of the enterprise, influencing the measured results or resulting in diagnosis error).

(b) Common or occasional nonconformity: Nonconformance of common or secondary indicators (unimportant quality characteristics).

4.2 Report

(a) Nonconforming products found in the product process shall be treated in accordance with *Deviation Treatment Management System*.

(b) Nonconforming products found in the inspection shall be recorded as specified by *Inspection Operating Procedure* in relevant inspection report, treated as per *Management System for Quality Inspection*, and the causes must be found and the problems solved if necessary.

4.3 Restrictive use / Using in "restrictive conformity" manner and return/replace-ment may be allowed for incoming noncon-forming products.

或 Incoming nonconforming products may be used in "restrictive conformity" manner, or returned or replaced.

4.3.1　Quality Control Department/QC shall provide Nonconformity Reports for nonconforming materials, which will be put in the nonconforming product area by Supply Department, and the return/replacement procedures shall be completed by the Procurement Center.

4.3.2　The decisions for using materials in "restrictive conformity" manner shall be made by QC, and if necessary, further verification shall be done by the Production Department or R&D Center with the Chief Engineer's approval.

4.4　The identification and treatment of nonconforming semi-finished and finished products. The treatment methods include rework and scrapping:

4.4.1　Common nonconforming products may be required to be reworked by production personnel immediately after being verified by PD and QC and made records for the inpsection on the record of semi-finished and finished products. The products that are judged as reworkable in subpackaging and assembling processes shall be reworked with the approval of the Quality Control Department.

(a) Rework shall be is completed by the Production Department, and reworked products shall be reinspected, with relevant inspection records be made.

(b) Scrapped products shall be put in noncon-forming product area by the Production Department and treated according to *Man-agement System for Nonconforming Product.*

4.4.2 Those products which are judged as seriously nonconforming shall be reviewed by the Production Director and approved by the General Manager, with signatures on relevant records for confirmation. The Quality Control Department shall treat nonconforming products as required and fill in the treatment records for them.

4.4.3　Relevant rework operating procedures shall be prepared for each rework, and reviewed by the Production Director or Chief Engineer for determining the influence of rework, and documented.

4.5　Review of reworked products

(a) The Technology Department, Quality Control Department and Production Department shall give suggestions on the possibility of rework, and predict the influence on production schedule caused by the disposal and corrective actions, and meanwhile, give advices on whether the applicability of raw materials or products and whether their influence on usage meet the requirement of laws and regulations.

(b) The Production Director or Chief Engineer shall organize related technical personnel to illuminate the nature about quality problems of raw materials, semi-finished and finished products.

(c) The Sales Department shall provide the market trend and offer proposals to the impact that may be caused by the rework.

(d) The Procurement Center shall provide information about the possibility and price of buying raw materials in time, and contact the suppliers so that relevant measures can be taken.

(e) The Nonconformities Report shall be filled out in accordance with the collective opinions by the Quality Control Department, stating the causes and opinions, submitted to the Production Director or Chief Engineer for approval, and notified to relevant departments for the organization and implementation. The Technology Department will prepare a rework process together with technical personnel if rework is needed, which shall be based on the *Management System for Rework*.

二、英译中

The maximum permitted concentrations in **non-exempt products**[①] are **0.1% or 1000 ppm** (except for cadmium, which is limited to 0.01% or 100 ppm) **by weight**[②]. The restrictions are on each **homogeneous material**[③] in the product, which means that the

---

[①] non-exempt products：非豁免产品。非豁免产品指的是 RoHS 中限定使用（不豁免）的物质。

[②] **0.1% or 1000 ppm by weight**：重量比 0.1% 或 1000ppm。

分析：原文的 0.1% or 1000 ppm 其实是相同意思，0.1% 就等于 1000 ppm，原文这样的表达，只是把同一含量的两种表达形式都写了出来，两个数字都不能省译。括号里的 0.01% 和 100ppm 也是同样的情况。

ppm：百万分之一。ppb：十亿分之一。两者均为浓度或概率的单位。

by weight：按重量（算），重量比。类似的表达还有：

by mass：按质量/重量（算），质量比/重量比。by volume：按体积（算），体积比。

weight (mass) / volumetric/ molar ratio：重量（质量）比/体积比/摩尔比。

[③] homogeneous material：均质材料（内部结构和组成完全均匀的材料）。

heterogeneous material：非均质材料（内部结构和组成不完全均匀的材料）。

limits do not **apply to**① the weight of the finished product, or even **to** a **component**②, but to any single substance that could (theoretically) be separated mechanically—for example, the sheath on a cable or the tinning on a component **lead**③.

As an example, a radio is composed of a **case**④, **screws**⑤, washers, a circuit board, speakers, etc. The screws, washers, and case may each be made of homogenous materials, but the other components comprise **multiple** sub-components of many different types of material⑥. For instance, a circuit board is composed of a bare **PCB, ICs**⑦, resistors, capacitors, switches, etc. A switch is composed of a case, a lever, a spring, contacts, pins, etc., each of which may be made of different materials. A contact might be composed of a copper strip with a surface coating. A speaker is composed of a permanent magnet, copper wire, paper, etc.

Everything that can **be identified as** a homogeneous material must meet the limit.⑧ So if **it turns out that** the case was made of plastic with 2,300 ppm (0.23%)

---

① apply to：适用于……注意：本句的 apply 后面有三个 to，即 apply to ... or even to ... but to ...
② component：组分；分量；部件/组件。
注意：component 在化学中经常指化学物质的"组分"，在力学中指力的"分量"，在机械设备中指机器的"部件"。
③ lead：*n.* 引导；导线。
④ case：*n.* 外壳；箱子。
⑤ screw：*n.* 螺钉；螺丝钉。注意："螺丝"是中国人对"螺丝钉"的口头称呼，不能用于书面语中。
⑥ The screws, washers, and case may each be made of homogenous materials, but the other components comprise multiple sub-components of many different types of material. 其中螺钉、垫圈和外壳可分别由均质材料制成，但其他部件则包含多个由许多不同类材料制成的次级部件。
分析：在本句中，multiple sub-components of many different types of material：多个由许多不同材料制成的次级部件，其中"multiple"的译文"多个"宜放在最前面，修饰的是"由许多不同材料制成的次级部件"，如果把"多个"直接放在"次级部件"之前，则会产生误解。
⑦ bare PCBs, ICs：裸板、集成电路。
PCB = printed circuit board：印制电路板（裸板）。注意：无特别说明时，PCB 都是指裸板。
IC = integrated circuit：集成电路。
说明：这两个缩写词都比较常见，都可以直接查网络词典得到其正确意思，一般都要译成中文。但要注意，ICs 是 IC 的复数，如果查 ICS 就很难查到，而查 IC 则可以比较快地查到。
⑧ **Everything** that can **be identified as** a homogeneous material must meet the limit. 所有可被认定为均质材料的部件均必须满足这种限制性要求。
everything：（意译）任何……的部件。be identified as：被认定为；被识别为。

PBB used as a flame retardant, then the entire radio would **fail** the requirements of the **directive**.①

In an effort to close **RoHS** 1 loopholes, in May 2006 the European Commission was asked to review two currently excluded product categories (monitoring and control equipment, and medical devices) for future inclusion in the products that must fall into RoHS compliance.② In addition the commission **entertains** requests for **deadline**

---

① So if **it turns out that** the case was made of plastic **with** 2,300 ppm (0.23%) PBB used as a flame retardant, then the entire radio would **fail** the requirements of the **directive**. 因此，如果可以证明外壳由塑料制成，其中含有 2,300ppm（2.3%）的 PBB 作为阻燃剂，则整台收音机都无法通过该指令的要求。

分析：本句中的 with 2,300 ppm (0.23%) PBB used as a flame retardant 是独立主格结构，表示伴随状态。另外，中文的连词搭配也要注意，如："如果……则……""除非……否则……""所有……都……"

it turns out that：结果证明；事实证明。turn out：证明；结果是。

fail：vt. 通不过（测试等）；达不到（要求等）；在……方面失败。

directive：n. 指令；命令。说明：欧盟关于环保方面的规定通常被称为 directive（指令）或 regulation（条例）。

② In an effort to close **RoHS 1** loopholes, in May 2006 the European Commission was asked to **review** two currently excluded product categories (monitoring and control equipment, and medical devices) for future inclusion in the products that must fall into **RoHS** compliance. 为堵住 RoHS 1 的漏洞，2006 年 5 月，欧洲委员会应邀对目前被排除在外的两个产品类别（监控设备与医疗器械）做出评审，以在将来将其列为必须遵守 RoHS 规定的产品。

分析：本句的主干是 the European Commission was asked to review ... for future inclusion。其中，review 的对象是 two currently excluded product categories，后面的 for ... 表示 review 的目的，而 inclusion 则与后面的 in 组成词组 inclusion in（include in 的名词形式，表示将某事物列入……）

RoHS 1 = RoHS 1.0：（不译）。

review：vt. 评估；评审。

RoHS = Restriction of the use of certain Hazardous Substances：（欧盟）限制在电子电气产品中使用有害物质的指令。

注意：这里的 RoHS 1 是指欧盟限制在电子电气产品中使用有害物质的指令的第一版，一般直接照抄不译，文中出现的 RoHS，一般也直接照抄。

extensions or for **exclusions by** substance categories, substance location or weight.① New legislation was published in the **official journal**② in July, 2011 which supersedes this exemption.

Note that batteries are not included within the scope of RoHS. However, in Europe, batteries are under the European Commission's 1991 Battery Directive (91/157/EEC), which was recently increased in scope and approved in the form of the new battery directive, **version 2003/0282 COD**, which will be official **when submitted to and published in** the EU's Official Journal.③ While the first **Battery Directive**④ **addressed**⑤ possible trade barrier issues brought about by **disparate**⑥ European member states' implementation, the new directive more explicitly high-lights improving

---

① In addition **the commission entertains** requests for **deadline extensions** or for **exclusions by** substance categories, substance location or weight. 此外，欧洲委员会还接受了按物质类别、物质位置或重量推迟限制使用的截止日期或排除限制的要求。

the commission = European Commission。entertain：*vt.* 接受；接纳。

deadline extension：期限延展。这里指将限制使用的截止日期往后推迟。

exclusion by substance categories, substance location or weight：按物质类别、物质位置或重量排除。exclusion：*n.* 排除；排斥。这里指将这些物质的使用排除在限制范围之外。by：按照。

② official journal = Official Journal of European Union：欧盟官方公报。

③ However, in Europe, batteries are under the European Commission's 1991 **Battery Directive** (91/157/EEC), which was recently increased in scope and approved in the form of the new battery directive, version 2003/0282 COD, which will **be official when submitted to and published in** the EU's Official Journal. 但是，在欧洲，电池已列入欧洲委员会 1991 年电池指令（91/157/EEC）中，最近这一指令的范围已经加大，并以新电池指令（即 2003/0282 COD 版指令）的形式通过，并将在提交欧盟官方公报并发布后正式生效。

分析：本句主干为 batteries are under the European Commission's 1991 Battery Directive (91/157/EEC)。句中第一个 which 引导的是 the European Commission's 1991 Battery Directive 的定语从句，version 2003/0282 COD 是 the new battery directive 的同位语。后一个 which 引导的是 the new battery directive 的定语从句，而 when submitted to and published in the EU's Official Journal 里的 submitted to 和 published in 的对象都是 the EU's Official Journal。

be official：正式生效。

when submitted to and published in ...：提交……并发布后。

④ Battery Directive：电池指令。

分析：由于这是 RoHS 的一份规范性文件，因此实词首字母全大写。其他实词首字母大写的词或词组，一般也是专有名词，前面的 EU's Official Journal 即是一例。

⑤ address：*vt.* 解决；处理。address 当"解决／处理"理解时，其宾语经常是 issue。

⑥ disparate：*adj.* 不同的；不一致的。这里宜将 disparate 译成"……的差异"。

and protecting the en-vironment from the negative effects of the waste contained in batteries. It also contains a programme for more ambitious recycling of industrial, automotive, and **consumer batteries**①, gradually increasing the rate of manufacturer-provided collection sites to 45% by 2016. It also sets limits of 5 ppm mercury and 20 ppm cadmium to batteries except those used in medical, emergency, or portable **power-tool devices**②. Though not setting **quantitative limits on quantities of**③ lead, lead-acid, nickel, and nickel-cadmium in batteries, it **cites**④ a need to restrict these substances and provide for⑤ recycling up to 75% of batteries with these substances. There are also provisions for marking the batteries with symbols **in regard to**⑥ metal content and recycling collection information.

  非豁免产品的以上成分最大许可浓度为重量比 0.1% 或 1000ppm（镉除外，其最大许可重量比浓度为 0.01% 或 100ppm）。这种限制是针对产品中的每种均质材料制定的，它意味着，这种限制不适用于成品重量，甚至也不适用于组件，而是只适用于（理论上）可被机械分离的任何单一物质——例如，电缆护套或元件引线的镀锡。

  举个例子，收音机由外壳、⑦ 螺钉、垫圈、电路板、喇叭等组成。其中螺钉、垫圈和外壳可分别由均质材料制成，但其他部件则包含多个由许多不同类材料制成的次级部件。例如，电路板就是由印制电路板（裸板）、集成电路、电阻器、电容器、开关等组成，而开关则由外壳、杠杆、弹簧、接触器、插脚等组成，而且他们各自还可分别由不同材料组成。其中，接触器可能由涂有表面涂层的铜带组成，而喇叭则由永久磁铁、铜线、纸等组成。

  所有可被认定为均质材料的部件均必须满足这种限制性要求。因此，如果可

---

① consumer battery：民用电池。
② power-tool device：动力工具设备；电动装置。
③ quantitative limits on quantities of ...：……的数量的定量限制。
④ cite：*vt.* 援引；引用。
⑤ provide for：规定。
⑥ in regard to：关于；与……相关。
⑦ 注意，本句译文的并列成分之间用的是顿号。当用于分隔两个或多个并列的小成分时，中文用顿号不用逗号。

以证明外壳由塑料制成，其中含有 2,300ppm（2.3%）的 PBB 作为阻燃剂，则整台收音机都无法通过该指令的要求。

为堵住 RoHS 1 的漏洞，2006 年 5 月，欧洲委员会应邀对目前被排除在外的两个产品类别（监控设备与医疗器械）做出评审，以在将来将其列为必须遵守 RoHS 规定的产品。此外，欧洲委员会还**接受**了按物质类别、物质位置或重量推迟限制使用的截止日期或排除限制的要求。2011 年 7 月欧盟官方公报发布的新法律取代 / 取消了这一豁免。

需指出的是，电池不在 RoHS 限制范围之列。但是，在欧洲，电池已列入欧洲委员会 1991 年电池指令（91/157/EEC）中，最近这一指令的范围已经加大，并以新电池指令（即 2003/0282 COD 版指令）的形式通过，并将在提交欧盟官方公报并发布后正式生效。尽管第一版电池指令解决了因欧盟成员国执行方面的差异而产生的可能的贸易壁垒问题，新指令更明确地强调了对环境免受电池所含废弃物的不良影响的改善和保护。该指令还包括了针对工业电池、汽车电池和民用电池的更宏大的回收计划，以将制造商提供回收点的比例，到 2016 年逐渐提高到 45%。该指令还为除医用、应急用或便携式电动装置用电池之外的其他电池，规定了汞 5ppm 和镉 20ppm 的限值。虽然该指令并未对电池中铅、铅酸、镍和镍镉的数量做出定量限制，但援引了限制这些物质的必要性，并规定含有此类物质的电池的回收率需达到 75%。此外，该指令还规定，电池上还必须标有与其金属含量和回收利用信息相关的符号。

第三节 机械工程制造

一、中译英
产品简介

本公司生产的<u>内胀式坡口机</u>，通过对管端内壁<u>胀紧</u>，经<u>行星减速器</u>多级<u>减速</u>后，<u>带动刀盘旋转</u>，为小口径管子<u>进行管端坡口</u>。适用管内径达到 **108mm**（如需坡口的管子口径超过此范围，可选购本公司的外钳式坡口机）。①

针对不同材料的管子②，只需安装对应材料的刀具即可安全快速加工，可按需指定不同角度坡口刀进行平口、坡口③。是您小口径管子加工的最佳选择。

---

① 本公司生产的内胀式坡口机，通过对管端内壁胀紧，经行星减速器多级减速后，带动刀盘旋转，为小口径管子进行管端坡口。适用管内径达到 108mm（如需坡口的管子①口径超过此范围，可选购本公司的外钳式坡口机）。Our ID-mounted bevellers can bulge the internal wall at the pipe end, and multi-stage decelerate by a planetary reducer, rotate the tool disc, and then bevel small-diameter pipes. The applicable inside diameters of pipes for our beveller reach 108 mm (if the inner-diameter of pipes to be beveled exceeds this scope, please choose our OD-mounted bevellers).

内胀式坡口机　ID-mounted beveller。外钳式坡口机：OD-mounted beveller。

说明：一般而言，中文"……机"，通常较少译成 XXX machine。"内胀式坡口机"这一名词不宜自己组合翻译，而是要从百度百科里查到相应的译文。在百度里查"内胀式坡口机 beveller"，可以得到 ID-mounted beveller 这一译法，从而得知其意思是为"内径安装式坡口机"。相应地，还可以得知"外钳式坡口机"的译文为 OD-mounted beveller，从而得知其意思为"外径安装式坡口机"。

行星减速器：planetary reducer。

说明：行星减速器，顾名思义就是像行星围绕恒星转动一样，有数个行星轮围绕一个太阳轮旋转的减速器。如下图：

胀紧：bulge / swell up。说明：胀紧是指通过机器的动作主动让管子膨胀，为"鼓起""凸起"之意。

减速：decelerate。说明：这里的减速是指自行减速，所以直接译成 decelerate。

带动……旋转：rotate ...

为……进行管端坡口 = 为……做坡口：bevel（vt.）。

刀盘 = 刀具盘：tool disc。刀具：tool。说明：刀具不译成 knife。

需坡口的管子：pipes to be beveled（后置定语）、to-be-beveled pipes（前置定语）。

小结：本句译文将使动词、被动语态或其他较长的动作表达法改成用主动语态的有 bulge、decelerate、rotate、bevel。另外，"需坡口的"译成 to-be-beveled，也省去了使用定语从句的变态译法。经过这样的改动之后，整个句子显得更简洁更生动。

② 不同材料的管子 = 由不同材料制成的管子：Pipes made from different materials 或 Pipes of different materials。

③ 指定不同角度坡口刀进行平口、坡口 = 用指定角度的坡口刀具（做）平口、坡口：be flushed or beveled with beveling tools of specified angles。

进行平口 = 做平口：flush（vt.）进行坡口 = 做坡口：bevel（vt.）。

BPP 系列**管道坡口机**①**广泛应用于**②**锅炉厂**③、压力容器、换热器、管道预制厂等**小口径管端平口、坡口需求领域**④。

特点

（1）**便携**⑤：我公司生产的 BPP 系列产品最大型号的长度和高度不超过 50cm，宽度不超过 20cm⑥，采用手提箱式包装，方便携带，使您可以轻松的在户外进行加工。

（2）**安装快速**：机器从箱体内取出，只需通过棘轮扳手使机器定位于管子中心并安装合适的刀具便完成准备工作。此过程不超过 3 分钟。按下电机按钮，机器便开始工作。

（3）**安全可靠**：**本产品常用的电动型配备麦太保角磨机**，通过角磨机内锥齿轮→行星齿轮减速器→主壳体内锥齿轮进行多级减速，在工作时输出低转速，

---

① 管道坡口机：beveller。说明：译成 beveling machine 太啰唆。
② 广泛应用于：be widely applied in、find wide applications in。
③ 锅炉厂：boiler plant。
　　plant 与 factory 的区别：plant 多指电力或机器制造业方面的工厂。factory 泛指一般意义的"工厂"，是最普通、用得最广泛的一个词。
④ 小口径管端平口、坡口需求领域：fields requiring flush or beveled small-diameter pipe ends。
⑤ 便携：Portability。
　说明：原文的"特点"部分共有四个小标题。在翻译时，前后并列的小标题一般要采用同样译法，而且小标题与文章标题一样，可以是非句子形式。这里将四个小标题都译成名词形式。
⑥ 我公司生产的 BPP 系列产品最大型号的长度和高度不超过 50cm，宽度不超过 20cm：our BPP series are not more than 50 cm in length and height and not more than 20 cm in width。
　　我公司生产的 BPP 系列产品 = 我们的 BPP 系列（产品）：Our BPP series。
　　最大型号的长度和高度不超过 50cm，宽度不超过 20cm = 最大长度和高度不超过 50cm，宽度不超过 20cm。

同时保留超大扭矩，使得坡口光滑平整，具有很高的质量，刀具寿命也因此加长。①

（4）**设计独特**：本产品主机采用航空铝材料，同时对各部分零件尺寸进行优化，使得机器小巧轻便。② **精心设计的**③胀紧机构让定位快捷准确，同时足够牢靠，使得加工时有足够高的刚性。产品配备各种样式的刀具使得机器能加工各种材料的管子以及各种角度的坡口、平口。④独特的结构以及自身润滑使得机器具有很长的寿命。

Product introduction

Our ID-mounted bevellers can bulge the internal wall at the pipe end, and

---

① 本产品常用的电动型配备麦太保角磨机，通过角磨机内锥齿轮→行星齿轮减速器→主壳体内锥齿轮 进行多级减速，在工作时 输出低转速， 同时保留超大扭矩，使得坡口光滑平整，具有很高的质量，刀具寿命也因此加长。Common electric bevellers are equipped with a Metabo angle grinder. Through multi-stage deceleration by an internal bevel gear of an angle grinder, a planetary reducer and an internal bevel gear of the main housing, the machines can operate under slow rotating speed while keeping extra large torque, produce high quality smooth and flat bevels, extending the service life of the tool.

  分析：本句主干为："本产品常用的电动型配备麦太保角磨机，通过……进行多级减速，（在工作时）输出低转速，同时保留超大扭矩"。其中，"通过……"是"进行多极减速"的方式状语，后面的"使得……"是前面动作的结果状语。

  本产品常用的电动型 = 常用的电动型坡口机：common electric bevellers。

  在工作时输出低转速 = 在低转速下工作：operate under slow rotating speed。

② 本产品主机采用航空铝材料，同时对各部分零件尺寸进行优化，使得机器小巧轻便：the machine is small and light since their main body is made from aviation aluminum and the sizes of all parts are optimized。

  分析：本句实际上是个因果句，前两小句是因，后一小句是果。如果顺着原文句式翻译，则难以译成好译文。而译成 the machine is small and light since ... and ... 应该是较好的译法。

  本产品主机采用航空铝材料 = 本产品由航空铝材料制成：the machine is made from aviation aluminum。

③ 精心设计的：well-designed。

④ 产品配备各种样式的刀具使得机器能加工各种材料的管子以及各种角度的坡口、平口。A variety of tools enable the machine to process pipes made from different materials and produce beveled ends with various angles and flush ends.

  分析：本句前面部分"产品配备各种样式的刀具"译文单独成句也可以，但句子会变得太长。所以，参考译文直接把这一部分当作主语，而且，将"产品配备"省译，这样的译文，意思跟原文也一样，但更简洁。另外，"各种角度的坡口和平口"宜理解成平口和各种角度的坡口，因为平口都是没角度的。

multi-stage decelerate by a planetary reducer, rotate the tool disc, and then bevel small-diameter pipes. The applicable inside diameters of pipes for our beveller reach 108 mm (if the inner-diameter of pipes to be beveled exceeds this scope, please choose our OD-mounted bevellers).

Pipes made from different materials can be safely and quickly processed by only installing tools of corresponding materials, and flushed or beveled with beveling tools of specified angles. Therefore, our products are your best choice for processing small-diameter pipes.

BPP series bevellers are widely applied (can find wide applications) in boiler plants, pressure vessels, heat exchangers, pipe prefabrication plants and other fields requiring flush or beveled small-diameter pipe ends.

Features

(1) **Portability**: our BPP series are not more than 50 cm in length and height and not more than 20 cm in width. They are packed in suitcases and portable, allowing you to process outdoors easily.

(2) **Quick installation**: after being taken out from the suitcase, the machine can be ready simply by positioning it at the center of the pipe with a ratchet wrench and equipping it with a suitable tool. The process will not exceed 3 min. The machine can start working by only pressing down the motor button.

(3) **Safety and reliability**: Common electric bevellers are equipped with a Metabo angle grinder. Through multi-stage deceleration by an internal bevel gear of an angle grinder, a planetary reducer and an internal bevel gear of the main housing, the machines can operate under slow rotating speed while keeping extra large torque, produce high quality smooth and flat bevels, extending the service life of the tool.

(4) **Unique design**: the machine is small and light since their main body is made from aviation aluminum and the sizes of all parts are optimized. The well designed expansion mechanism can realize quick and precise positioning, moreover, it is strong enough, with sufficient processing rigidity. A variety of tools enable the machine to process pipes made from different materials and produce beveled ends with various angles and flush ends. Moreover, its unique structure and self-lubrication function

endow the machine with rather long service life.

二、英译中

In another study, the effect of the **microchannel condenser**[①] on a **split-ducted cooling system's**[②] performance was analyzed. A commercially available system based on a copper **RTPF condenser**[③], and R410a refrigerant was experimentally evaluated first to establish a **baseline**[④]. Then, the original condenser was replaced with a microchannel heat exchanger of almost identical external volume, **face area**[⑤], and **fin pitch**[⑥]. The results of this study indicated that the microchannel condenser improves **COP**[⑦], condenser capacity, and evaporator capacity, when compared to the standard baseline system. These are caused by the superior heat transfer characteristics of the microchannel condenser, lower refrigerant-side pressure drop, and consequently lower condensing temperature – resulting in a reduction in the amount of **work**[⑧] needed from the compressor. All of this acts to increase both the system's capacity and COP. The air-velocity distribution on the surface of the two condensers was similar, and that

---

① microchannel condenser：微通道冷凝器。
    说明：在百度里查"microchannel condenser 冷凝器"，即可得到这一产品的中文名称。
② split-ducted cooling system：分体式管道冷却系统。split：*adj.* 分体（式）的。
    说明：在百度里查"split-ducted cooling system 冷却系统"，可得到 Ducted Split Cooling System，再查"Ducted Split Cooling System 冷却系统"，可得到相关名词"分体式管道酒窖冷却系统"，因此，将 split-ducted cooling system 译成"分体式管道冷却系统"。
③ RTPF condenser = round-tube, plate-fin condenser：圆管板翅式冷凝器。
    说明：这一产品的英文，可以在必应（Bing）里查"RTPF condenser"，查到全称 round-tube, plate-fin condenser。查到全称后，再在百度里查"round-tube, plate-fin condenser 冷凝器"，即可得到其中文名称。
④ baseline：*n.* 基准；基线。
⑤ face area：（制冷）迎风面积。
⑥ fin pitch：翅片距离。pitch：*n.* 距离。
⑦ COP = Coefficient of Performance：（热力学）制冷系数。
    分析：COP 的全称，可以在必应（Bing）里查"COP stands for"，然后打开相关链接，从 COP 的一系列缩写中，发现最适宜的一个为 Coefficient of Performance，然后，再在网络词典里查 Coefficient of Performance 得到其中文意思。需注意的是，COP 是制冷中非常常见的一个参数，所以，如果不是在公式或表达式里，一般要译成中文。
⑧ work：*n.*（力学、热力学）功。

distribution is related to the distance between the condenser fan and the condenser surface. The dead zones of air-flow were measured in corners of the condensers and in the folded parts. The air-side pressure drop distribution matched well with the distribution of **velocity**①. In this study, the value of U_{air} decreased as **subcooling**② increased; this phenomenon is physically valid. The higher thermal performance, per unit base surface area A, of **MPE** heat exchangers may be used to achieve one of the three objectives below:③

· Heat Exchanger and Overall Equipment Size Reduction: If the heat exchange rate (Q) is held constant the heat exchanger length may be reduced. This will provide a smaller heat exchanger.

· **Increased heat exchange rate**④: This may be exploited either of two ways:

— Reduced driving temperature difference (Δ Tm): If Q and the total tube length (L) are held constant, the Δ Tm may be reduced. This provides increased thermodynamic process efficiency, and **yields**⑤ a savings of operating costs to the end user.

— Increased Heat Exchange Rate: Keeping L constant, the increased UA/L will

---

① velocity = air-velocity。

② subcooling: *n.* 过冷（度）；再冷却。

③ The higher thermal performance, per unit base surface area A, of **MPE** heat exchangers may be used to achieve one of the three objectives below 可以通过 MPE 换热器较高的单位底面面积 A 的热性能来实现下列三个目标中的任意目标。

　　分析：本句中间的 per unit base surface area A 是 The higher thermal performance 的后置定语，这一定语插入在 The higher thermal performance 和另一定语 of MPE heat exchangers 之间，加大了理解难度。翻译时，宜将两个定语的译文都放在被修饰词的译文前，而且，per unit base surface area A 的译文应该直接放在被修饰词的译文前。

　　MPE = multiport extrusion：多端口挤压（本节选部分的前文出现过，这里不译）。

④ Increased heat exchange rate: 提高热交换率。

increased: *adj.* 提高的；更高的。说明：本文 increased、reduced、fixed 等都是以动词过去分词当形容词的例子。

⑤ yield: *vt.* 产生；产出。

result in increased heat exchange rate for **fixed fluid inlet temperatures**①.

This increase in **heat transfer capacity per unit volume** has been confirmed experimentally for a variety of refrigerants and specific HVAC&R applications.②

Brazed③ aluminium MPE heat exch-angers have been implemented commercially in residential and commercial HVAC products by York (JCI) and Carrier. These manufacturers have been reaping the benefits of this conversion for some time as well as providing confirmation of these benefits on a large commercial scale. They are also well positioned to react to the ever-changing regulatory landscape. MPE tubes are exceptionally capable of handling high pressure refrigerants, and the process of making MPE makes it easy to adapt tube **geometry** to meet a wide range of refrigerant operating pressure conditions.④ Whether the **driver**⑤ for refrigerant changes comes from technical requirements, consumer demands or regulatory pressures, the MPE concept can be easily adapted to meet those needs.

---

① Keeping L constant, the increased UA/L will result in increased heat exchange rate for **fixed fluid inlet temperatures**. 使 L 保持恒定，则在液体进口温度固定不变的情况下，UA/L 的增大将会导致热交换率升高。

分析：本句后面的 for fixed fluid inlet temperatures 的意思，是指"对于固定的液体进口温度"，即"在液体进口温度固定不变的情况下"。

fixed fluid inlet temperature: 固定的液体进口温度；液体进口温度固定不变。fixed: *adj.* 固定的。

② This increase in **heat transfer capacity per unit volume** has been confirmed experimentally for a variety of refrigerants and specific **HVAC&R** applications. 对各种制冷剂和具体暖通空调与制冷应用，也已经通过实验确认单位容积传热能力的提高。

heat transfer capacity per unit volume: 单位容积传热能力。

HVAC&R = heating, ventilation, air conditioning and refrigeration：暖通空调与制冷。

说明：HVAC&R 的全称，可通过在必应（Bing）里查"HVAC&R stands for"查到。其中 HVAC 是制冷方面非常常见的缩写，可以当作一个单词记住。

③ brazed：*adj.* 钎焊的。

④ MPE tubes **are exceptionally capable of** handling high pressure refrigerants, and **the process of making MPE** makes it easy to adapt tube **geometry** to meet a wide range of refrigerant operating pressure conditions. MPE 管处理高压制冷剂的能力特别突出，MPE 的制作工艺，使得通过更改管形以满足各种制冷剂操作压力条件变得更加容易。

be exceptionally capable of: 特别能；在……方面能力特别突出。

the process of making MPE：MPE 的制作工艺。

geometry：*n.* 几何图形。这里将其理解为"外形"。

⑤ driver：*n.* 动因；驱动因素。

在另一项研究中，则对微通道冷凝器对分体式冷却系统性能的影响进行了分析。首先对以铜管圆管板翅式冷凝器和 R410a 制冷剂为基础的商用系统进行了实验性评估，以便确立基准。然后，用外部体积、迎风面积和翅片间距几乎都相同的微通道换热器替换了原冷凝器。此项研究结果表明，与标准基准系统相比，微通道冷凝器的制冷系数、冷凝器容量和蒸发器容量都提高了。这是因为微通道冷凝器具有较高的传热特性和较低的制冷剂侧压降，从而得到较低的冷凝温度，最终降低压缩机所需做功量。所有这些原因一起作用，就会提高系统的容量和制冷系数。两种冷凝器表面的空气速度分布相似，这种分布形式与冷凝器风机和冷凝器表面之间的距离有关。气流死区在冷凝器各个角落和折叠部位测量。空气侧压降分布与空气速度分布吻合。在此项研究中，U 空气值随着过冷度的升高而降低；从物理学角度说，这种现象是合理的。可以通过 MPE 换热器较高的单位底面面积 A 的热性能来实现下列三个目标中的任意一个：

· 换热器和设备外形尺寸减小：如果热交换率（Q）保持恒定，则换热器长度可以缩短。这样则可准备一台较小的换热器。

· 热交换率增大：可以通过下列两种方法中的任意一个实现：

—— 降低驱动温差（ΔT_m）：如果 Q 和总管长（L）保持恒定，则 ΔT_m 可能会降低。这可提高热力学过程效率，并为终端用户节省操作成本。

—— 提高热交换率：使 L 保持恒定，则在液体进口温度固定不变的情况下，UA/L 的增大将会导致热交换率升高。

对各种制冷剂和具体暖通空调与制冷应用，也已经通过实验确认单位容积传热能力的提高。

York（JCI）和 Carrier 已经在家庭和商业暖通空调产品中使用了钎焊铝 MPE 换热器。在一段时间内，这些制造商一直在获取这种转变带来的好处，并以商业大规模生产证实这些好处。他们在对不断变化的监管环境做出反应方面也处于有利地位。MPE 管处理高压制冷剂的能力特别突出，MPE 的制作工艺，使得通过更改管形以满足各种制冷剂操作压力条件变得更加容易。无论制冷剂更换的原因是来自技术要求、客户要求，还是监管压力，MPE 概念都能轻易地适应这些要求。

第四节 物理、电子电工

一、中译英

操作方法：

（1）准备：

a、上电①。上电前请先按电路图确认接线正确无误，确认无误后，将操作面板上的"急停"开关按下，并将手动/自动选择开关置于手动位置，方可合上总电源开关 QF1。②然后合上控制电源开关 QF2，关好柜门后，再右旋面板上的"急停"开关，使其弹出复位。③

---

① 上电：power on。说明：上电＝通电；下电＝断电。

② 上电前请先按电路图确认接线正确无误，确认无误后，将操作面板上的"急停"开关按下，并将手动/自动选择开关置于手动位置，方可合上总电源开关 QF1。Before powering on, check the wiring by circuit diagram, then press the "Emergency Stop" button on the operation panel, and turn Manual/Auto selector switch to manual position, and then main switch QF1 can be closed.

分析：本段（包括本句在内）使用的都是典型的说明书句式，即祈使句。祈使句一般是表示要求、指示、命令等的句式，其动作尚未发生，因此不宜译成主动句，也不宜译成被动句，而是译成跟中文原文一样的祈使句。本文件的很大一部分内容都使用这种句式。

电路图：circuit diagram。说明：diagram 指的是以简单图线框对产品或过程所做的图示和解释（结构/功能/逻辑/过程的可视化）。

确认接线准确无误：check the wiring。确认＝检查/查核：check。接线：wiring。

确认无误后：（啰唆冗余部分，省译）。面板：panel。

急停：Emergency Stop；E-Stop。注意：表示按钮名称时，其英文首字母要大写。

手动/自动 Manual/Auto。选择开关 selector switch（一般不译成 selection switch）。

合上/断开 close/open。

注意：合上开关（close the switch）＝通电（turn on the switch）；断开开关（open the switch）＝断电（turn of the switch）。而且，"打开/关上开关"的说法不规范，不能直接译成 Open/close the switch，否则会译反。

③ 然后合上控制电源开关 QF2，关好**柜门**后，**再右旋**面板上的"急停"开关，使其弹出复位。After that, close control power switch QF2, and close the cabinet door before rotating "Emergency Stop" switch clockwise on the panel for popup and resetting.

分析："关好柜门后，再右旋面板上的'急停'开关"一般可译成 close the cabinet door, then rotate the "Emergency Stop" switch clockwise on the panel for popup，但这里将 then rotate 改成 before rotating，也就是说，将"在 A 动作后做 B 动作"理解成"在 B 动作之前做 A 动作"，译成 A before B。这是正说反译的方法，这种译法显得更灵活。

柜门＝电柜门：cabinet door。电柜：electric cabinet。这里简译为 cabinet。

右旋＝顺时针旋转：rotate clockwise。

b、将顶板选择开关置于<u>左边向上位</u>，<u>使</u>顶板机构<u>动作</u>，再放入<u>要加工的工件（板材）</u>，移动板材到合适位置。^① 然后<u>调好</u>^② <u>左右限位开关</u>^③的位置（接近传感器），开关动作时，切割圆盘必须<u>移出</u>^④板材头部或尾部。并且将切割圆盘<u>行走</u>^⑤到右限位。

c、将顶板选择开关置于右位向下，<u>使顶板装置下降</u>^⑥，将板材放置到台板上，再将压板选择开关置于右向下位，使压板气缸下压压住板材。准备工作完毕。

注意：顶板机构未降下时，压板机构不能下压。顶板升起时压板机构同时升起。<u>顶板机构未降，下压板机构未下压，不能自动工作</u>^⑦。切割圆盘位于左限位时，圆盘不能启动。<u>开始切割工作之前</u>^⑧，一定要将切割圆盘行走到右限位处。

---

① 将<u>顶板</u>选择开关置于<u>左边向上位</u>，<u>使</u>顶板机构<u>动作</u>，再放入<u>要加工的工件（板材）</u>，移动板材到合适位置。Place the jack plate selector switch on the left upward position to actuate the jack plate mechanism, and then put in the workpieces (plates) to be processed, and move the plates to the appropriate position.

　　顶板：jack plate。压板：press plate。
　　注意：勿将"顶板"译成 roofplate（建筑的"屋面板"），而是要译成 jack plate（机械的"顶升板"）。因为后文还有与"顶板"对应的"压板"（译为 press plate），"顶"和"压"刚好相反，但 roof 和 press 却不相反。因此，可推测"顶板"应改用其他译法。而 jacking 是顶升的意思，但在必应（Bing）里搜索 jacking plate，显示的却基本上是 jack plate，而且 jack plate 的英英解释也是"顶板"之意，因此，jack plate 才是"顶板"的正确译法。
　　左边向上位：left upward position。
　　使……动作：actuate；drive。
　　要加工的工件（板材）：workpieces (plates) to be processed 或 to-be-processed workpiece (plates)。

② 调好 = 调节好 = 调节：regulate。
　　区别：regulate（调节）是指对机械设备等的按键或参数（如大小、尺寸）等来回改动，直到达到理想情况。而 adjust（调整）则是指对某些情况进行变更，即从一种不合理或不想要的情况变成另一种情况。

③ 限位开关：limit switch。
④ 移出：move out of。
⑤ 行走 = 移动：travel 或 move。说明：travel 经常用于指机械设备的"行走"。
⑥ 使……下降：descend。说明：这里的"下降"不宜译为 fall。
⑦ 顶板机构未降，下压板机构未下压，不能自动工作。Automatic operation is not allowed before des-cending the jack plate mechanism or pressing down the press plate mechanism.
　　说明：本句的"不能自动工作"是指不允许或者无法自动工作，但不明确到底是不允许还是无法，但不论是不允许还是无法，都可以译成 is not allowed，所以不译成 is impossible。

⑧ 开始切割工作之前……：... before cutting。

(2) 手动工作：

经以上准备工作后。①将操作面板上的"手动/自动"选择开关，向右旋到手动位置，按下"切割圆盘启动"按钮待切割圆盘（**切刀**②）启动后，按住**"向左或向右切割"按钮**③进行向左或向右切割。

(3) 自动工作：

经以上准备工作后。先按下"切割圆盘启动"按钮待切割圆盘（切刀），启动后，再将操作面板上的"手动/自动"选择开关向左旋到自动位置，切割圆盘自动行走到左限位停止，并且圆盘旋转也停止。再次自动工作时，必须以手动将切割圆盘行走到右限位方能工作。

操作注意事项：

(1) 当变频器发生**故障**④时，变频器操作面板上也会显示**当前故障代码**⑤，请**查阅**⑥变频器说明书，了解发生该故障的**原因及排除方法**⑦。按下变频器故障复位键之前，一定要检查故障原因，如果找不到故障原因，在确认机器安全后，可按下变频器复位键，如按复位键后故障代码**不能消失**，可**断开**总电源开关 2 分

---

① 经以上准备工作后。After the preparation above...

分析：这一小句后面的句号用法不妥，对于这种情况，在英译时，宜将其视为逗号，将本句与后面部分内容合并翻译。在实际应用类翻译中，经常会出现原文断句不合理现象，或者说虽然原文断句合理，但英译时按照原文断句却不妥当的情况，这种情况下，译文断句一般要根据实际需要确定，而不是完全跟从原文断句。

② 切刀：cutter。说明：机械设备的"刀"一般不译成 knife，因为 knife 通常指日常所用的小刀。

③ "向左或向右切割"按钮："Left Cut" or "Right Cut" (button)。

分析：这部分原文可能有错，正确表达应该是"'向左切割'或'向右切割'按钮"，所以译成 "Left Cut" or "Right Cut" (button)，其中 button 可省译。如果译成 "Cutting Leftward" or "Cutting Rightward" button 则显得啰唆。

④ 故障：trouble；fault；failure。其中 trouble 更经常用于指某个具体故障。

⑤ 当前故障代码：present trouble code。

分析：这里的"当前"宜译成 present，不宜译成 current，因为 current 还有"电流"的意思，易导致误解。

⑥ 查阅（说明书等）：refer to。

⑦ 原因及排除方法：causes and countermeasures。排除方法 = 对策：countermeasure。

钟后上电再试。① 如仍不能排除，可能的原因是变频器损坏。

（2）按压按钮开关时严禁**野蛮操作**②，宜稍用力向下按压到位后迅速放开。

（3）当发生紧急情况按下急停按钮会使整机立即停止工作。③松开急停按钮之前请确认安全。

（4）器件损坏更换时首选与原品牌型号相同的器件，如果采用其他品牌代换，一定要用型号、规格相同的器件更换，否则将会造成更大的设备事故。④

（5）做好防尘散热工作⑤，对柜内积尘定期清理，疏通风道，保证通风良好。

Operation method:
(1) Preparation:
a. Power on. Before powering on, check the wiring by circuit diagram, then press

---

① 如按复位键后故障代码**不能消失**，**可断开**总电源开关 2 分钟后上电再试。if the trouble code still exists, disconnect the main power switch for 2 minutes then power on to have another try.

  分析：本句的"不能消失"译成 still exist，是正说反译。这样的例子在前文也出现过一次（即"在关好柜门后，再右旋面板上的'急停'开关，使其弹出复位"）。

  断开（连接、电源等）：disconnect。

② 野蛮操作 = 用过很重的操作：heavy operation。

③ 当发生紧急情况按下急停按钮会使整机立即停止工作 = 在紧急情况下，按下急停按钮，让整体立即停止运行。Under emergent conditions, press emergency stop button to stop the whole machine immediately.

④ 器件损坏**更换**时**首选**与原品牌**型号**相同的器件，如果采用其他品牌代换，一定要用型号、**规格**相同的器件更换，否则将会造成更大的设备事故。

  译一：When parts are damaged and need to be replaced, the parts of the same models as original parts are preferred, if the parts of other brands are used, they must be parts of the same models and types, otherwise more serious accident will be caused.

  译二：The models same as the original parts are preferred when replacing damaged parts, if not, parts with the same types and parameters shall be adopted, otherwise a more serious trouble for the equipment may be caused.

  分析：以上译二为参考译文，译一为顺着原文思路得出的译文。很明显，译一不如译二顺畅，或者说有点中式，英语译二是经过转换说法后得到的译文。另外，中文里常见的"规格"，在这里指的是类型之类的意思。

  更换：replace。

  区别：replace（替换；更换）是指用 B 替换、取代 A，而 change（改变；变化）则是指发生变化，change 表示更换时，一般用于更换服饰。

  首先：be preferred。型号：model。

⑤ 做好防尘散热工作 = 确保防尘和散热良好：Ensure good dedusting and heat dissipation。

the "Emergency Stop" button on the operation panel, and turn Manual/Auto selector switch to manual position, and then main switch QF1 can be closed. After that, close control power switch QF2, and close the cabinet door before rotating "Emergency Stop" switch clockwise on the panel for popup and resetting.

b. Place the jack plate selector switch on the left upward position to actuate the jack plate mechanism, and then put in the workpiece (plates) to be processed, and move the plates to the appropriate position. Regulate the position of left and right limit switch (close to a sensor). When the switch moves, the cutting disc has to move out of the head or rear of plate. Travel the cutting disc to the right limit.

c. Place the jack plate selector switch on the right downward position to descend jack plate device and put the plate on bedplate; and place the press plate selector switch on the right downward position to descend press plate cylinder and press the plate. And the preparation work is completed.

Note: Press plate mechanism should not press down before descending the jack plate mechanism. Jack plate should rise together with the press plate mechanism. Automatic operation is not allowed before descending the jack plate mechanism or pressing down the press plate mechanism. The disc must not start when cutting disc is on the left limit. The cutting disc must be travelled to the right limit before cutting.

(2) Manual operation:

After the preparation above, turn the "Manual/Auto" selector switch on operation panel to the manual position on the right, press down "Start Cutting Disc" button; after starting the cutting disc (cutter), press the button of "Left Cut" or "Right Cut" to cut leftward or rightward.

(3) Automatic operation:

After the preparation above, Press "Start Cutting Disc" button first, after starting the cutting disc (cutter), rotate the "Manual/Auto" selector switch on operation panel to the automatic position on the left, the cutting disc will automatically travel to the left limit and stop, and the disc also stop rotating. Cutting disc must manually travel to the right limit for the next automatic operation.

Operating precautions:

(1) When any trouble occurs to the converter, the operation panel of converter will show present trouble code. Please refer to converter manual for the causes and countermeasures of the trouble. Trouble causes must be checked before pressing the trouble reset button of converter. Where no trouble cause can be found and the machine is confirmed to be safe, the converter reset button can be pushed, if the trouble code still exists, disconnect the main power switch for 2 minutes then power on to have another try. Where the trouble cannot be removed, the converter might breakdown.

(2) Heavy operation is forbidden when pressing the button, it is recommended to push down to its position and set free quickly.

(3) Under emergent conditions, press emergency stop button to stop the whole machine immediately. Safety must be ensured before loosening the emergency stop button.

(4) The models same as the original parts are preferred when replacing damaged parts, if not, parts with the same types and parameters shall be adopted, otherwise a more serious trouble for the equipment may be caused.

(5) Ensure good dedusting and heat dissi-pation, clear the dust in the cabinet regularly, smooth the air ducts for a good ventilation.

二、英译中

Dielectric Frequency Response (DFR)[①] testing of transformers is increasingly **acknowledged**[②] as the most dependable and most convenient way of determining the condition of the insulation in power transformers and **bushings**[③]. Crucial benefits of DFR testing are that the tests can be carried out at any temperature, it measures **tan delta**[④] and capacitance, it is able to evaluate the condition of the insulation, not only of moisture content in the paper insulation, but also the dielectric properties of the oil and finally it can determine the temperature **dependence**[⑤] in the transformer or bushing.

---

① Dielectric Frequency Response (DFR)：介电频率响应。
② acknowledge：*vt.* 认为；公认；确认。
③ bushing：*n.* 套管。说明：这里的 bushing，指的是 cable bushing（电缆套管）。
④ tan delta：（介电损耗）角正切。
⑤ dependence：*n.* 相关性；依赖性。说明：相关性指的是两个变量的关联程度。

All in one single test!①

Because of this, DFR testing gives a much more comprehensive and dependable indication of the transfor-mer's **overall health**② than other test techniques. In addition, with appropriate equipment, tests take only a relatively short time to complete.

Useful as the DFR testing technique undoubtedly is, however, it can sometimes be difficult to use successfully in substations where there is an extremely high level of **electrical noise**, e.g. during high humidity or in **HVDC** converter stations.③ This is hardly surprising, as standard instruments carry out DFR tests with an applied voltage in the region of 200 V (peak). This is perfectly adequate for most test objects and environments, but in harsh environments, even with the advanced **filtering**④ technique in IDAX300, it may not be possible to measure the full frequency range with adequate accuracy.

A recent development has, however, provided a complete and effective solution to this problem. It sounds **deceptively** simple – an amplifier to increase the output voltage

---

① All in one single test! 而且都只需测试一次！
分析：这里的 all 指的是以上所述的所有评估，句子的意思是所有评估都只需要一次测试便可完成。

② overall health：整体健康状况；整体状况。

③ Useful as the DFR testing technique undoubtedly is, however, it can sometimes be difficult to use successfully in substations where there is an extremely high level of **electrical noise**, e.g. during high humidity or in **HVDC** converter stations. 毫无疑问，介电频率响应测试技术是很有用的。然而，在变电站电气噪音水平极高的情况下，介电频率响应测试有时也很难成功运用，比如：在高湿度条件或者高压直流变电站中。
分析：本句主干是 it can sometimes be difficult to use successfully in substations。开头（Useful as the DFR testing technique undoubtedly is）是一个倒装的让步状语从句，后面 where 引导的是 substations 的地点状语从句。
electric noise：电气噪音。HVDC = high voltage direct current：高压直流（可在网络词典里查全称）。

④ filtering：n. 滤波。

of a DFR **test** set by a **factor** of ten to 2 kV.① Unsurprisingly, actually producing a small and efficient amplifier that will operate accurately and reliably is technically challenging, but this has now been achieved.

The increased output voltage, combined with advanced digital filtering algorithms greatly reduces the influence of electromagnetic interference, allowing accurate and dependable results to be obtained under conditions where this would previously have been difficult or even impossible②.

The best DFR test sets are also surprisingly fast in operation. With efficient signal **acquisition**③ technology and two separate current measurement channels, these instruments provide reliable true **AC**④ measurements without the need for **DC**⑤ to AC conversion. This means, for example, that two separate measurements down to 1 **mHz** on a three winding transformer can be completed in a little over half an hour, whereas

---

① It sounds **deceptively** simple – an **amplifier** to increase the output voltage of a DFR test set **by a factor of ten** to 2 kV. 这一方案似乎很简单——只需使用放大器将介电频率响应测试设备的输出电压放大十倍至2kV。

分析：本句的理解难点有两处：一是 deceptively，其本意是"迷惑(性)地"，这里宜理解为"似乎"。二是 by a factor of ten，其意思是"以十倍的系数"，increase by a factor of ten 的意思是"放大十倍"。

deceptively：*adv.* 迷惑(性)地；欺骗性地。test set：测试设备。
amplifier：*n.* 放大器。factor：*n.* 系数；因数。

② ... allowing accurate and dependable results to be obtained under conditions where this would previously have been difficult or even impossible. 使得在之前被认为极其困难甚至是不可能的条件下，也能获得精确可靠的测试结果。

分析：本部分内容是主句的伴随状态，其翻译难点在于句子没有施动者，如果在译文加入一个施动者（如"测试人员"），译成"……因此就能够让测试人员即使在先前很困难或者甚至是不可能的那些条件下，也能获得精确可靠的测试结果"，似乎句子也很顺。但更合适的译法，是不增译，将 this would previously have been difficult or even impossible 译成"在之前被认为极其困难甚至是不可能的情况下"，这样句子更为简洁。

③ acquisition：*n.* （信号、资料等的）采集；收集。
④ AC= alternating current：交流；交流电。
⑤ DC = directive current：直流；直流电。

instruments of more conventional design typically take around two hours.①

It is also worth noting that the use of a high-voltage amplifier with a suitable DFR test set extends the capabilities of the instrument to include **power frequency** (50/60 Hz) capacitance and tan delta measurements at higher voltage.② The maximum capacitance of a test object that a test set can handle will depend on the current output capability and VAX020, with an amplifier capable of supplying 50 mA, can test objects

---

① This means ,for example, that two separate measurements down to 1 **mHz** on a three winding transformer can be completed in a little over half an hour , whereas **instruments of more conventional design typically** take around two hours. 举例而言，这就意味着，能够在半个小时多一点点的时间内，完成三绕组变压器上两个低至 1mHz 的独立测量，而常规设计的旧仪器完成同样任务通常需要约两个小时的时间。

    分析：本句主干为 This means that two separate measurements can be completed in a little over half an hour，其中 for example 是主句谓语和宾语之间的插入语，理解时可以跳过。down to 1 mHz on a three winding transformer 是 measurements 的后置定语，后面的 whereas 则是转折状语从句。

    mHz：毫赫兹。

    说明：这里的 mHz 只有 H 是大写的，MHz 或 Mhz 都是不规范写法。其中 m 是"毫"的意思，1mHz 相当于 1000 秒才一个周期，是很低的频率。

    instruments of more conventional design：更常规设备的仪器。这里意译为常规设计的旧仪器。

    typically：adv. 典型地；有代表性地；通常地。说明：在很多情况下，"有代表性地"其实就是"通常地"。

② It is also worth noting that the use of a high-voltage amplifier with a suitable DFR test set extends the capabilities of the instrument to include **power frequency** (50/60 Hz) capacitance and tan delta measurements at higher voltage. 同样值得注意的是，具有合适介电频率响应测试设备的高压放大器，可将设备的测量能力扩展至在更高电压下的工频（50/60Hz）电容和损耗角正切。

    分析：本句的结构是 It is also worth noting that ... 句子重心是 that 引导的宾语从句。该从句结构为 the use of a high-voltage amplifier (with ...) extends the capabilities (of the instrument to include ... and ... at ...)，其意思是使用具有合适介电频率响应测试设备的高压放大器可以扩展其测量能力，to include 后的内容，说明的就是其测量能力可扩展到什么程度。

    power frequency：工频。

    背景知识：工频（power frequency）是指电力系统的发电、输电、变电与配电设备以及工业与民用电气设备采用的额定频率，单位赫兹（Hz）。中国的工频采用 50Hz，有些国家采用 60Hz。所以这里在 power frequency 后面的括号里加注"50/60Hz"。

of up to 80 nF at 50 Hz and 67 nF at 60 Hz.① Besides testing transformers and bushings this is also a great test setup for measuring tan delta and/or conductivity of insulating oil where the international standards typically recommends a test voltage of 200 to 1000 V/mm.

An excellent example of a high-voltage amplifier and DFR test set combination is Megger's VAX020 amplifier used in conjunction with the company's popular and efficient IDAX300 tester. This combination with a total weight of less than 10 kg, reliably and cost-effectively extends the invaluable technique of DFR testing into areas where its application was formerly difficult or impossible.②

变压器的介电频率响应测试正越来越被认为是确定电力变压器和套管中绝缘情况的最可靠、最便捷的方式。介电频率响应测试的关键优势在于：可在任何温度下进行测试，可测量损耗角正切和电容，能够评估绝缘情况，不仅仅能评估纸绝缘的水分含量，还能评估油的介电特性。最后本测试还能够确定变压器或套管中的温度相关性。而且都只需测试一次！

正因如此，跟其他测试技术相比，介电频率响应测试能够更全面可靠地显示

---

① The maximum capacitance of a test object that a test set can handle will depend on the current output capability and VAX020, with an amplifier capable of supplying 50 mA, can test objects of up to 80 nF at 50 Hz and 67 nF at 60 Hz. 测试设备能够测量的测试对象的最大电容取决于放大器的电流输出能力，而 VAX020 具有能供应 50mA 电流的放大器，则可对 50Hz 下电容高达 80nF 或 60Hz 下电容高达 67nF 的测试对象进行测试。

　　分析：本句有两个句子。第一句主干为 The maximum capacitance of a test object (that ...) will depend on the current output capability，其中 that 引导的是 a test object 的定语从句。第二句主干为 VAX020 can test objects (of ...)，其中 with 引导的是主语 VAX020 的后置定语，这一后置定语（with ...）插入在主谓之间，增加了句子的理解难度。

② This combination with a total weight of less than 10 kg, reliably and cost-effectively extends the invaluable technique of DFR testing into areas where its application was formerly difficult or impossible. 这种组合的重量不超过 10kg，将介电频率响应测试技术这一极有用的技术，可靠且低成本地拓展到之前很难甚至是不可能的应用中。

　　分析：本句主干为 This combination extends the invaluable technique of DFR testing，其中 with a total weight of less than 10 kg 是 This combination 的后置定语，into areas where ... 是 extend 的间接宾语。其中 this combination = a high-voltage amplifier and DFR test set combination。由于原文结构较为复杂，句子较长，在组织译文时，宜在句子的适当位置加入逗号，表示语气停顿。

出变压器的整体状况。另外，若有合适设备，则介电频率响应测试只需花费较短时间即可完成。

毫无疑问，介电频率响应测试技术是很有用的。然而，在变电站电气噪音水平极高的情况下，介电频率响应测试有时也很难成功运用，比如：在高湿度条件或者高压直流变电站中。这并不奇怪，因为标准设备进行介电频率响应测试时所用电压都是在200V（峰值）区域中。这对于大部分的测试对象和测试环境来说已经完全足够了；但在某些恶劣环境中，即使使用IDAX300的先进滤波技术，本测试也不可能以足够的准确度测量整个频率范围。

然而，针对这一问题，最近研发出了一种完整有效的解决方案。这一方案似乎很简单——只需使用放大器将介电频率响应测试设备的输出电压放大十倍至2kV。然而，实际上生产一款合适的、可精确可靠地操作的放大器却是极具技术挑战性的。但现在这个难关已被攻克。

输出电压的提高，再加上先进的数字滤波算法的运用，可大大减少电磁干扰的影响，使得在之前被认为极其困难甚至是不可能的条件下，也能获得精确可靠的测试结果。

最好的介电频率响应测试设备的运行也极其快速。这些设备运用高效的信号采集技术，具备两个独立的电流测量通道，能够直接提供可靠的真实交流测量值，且无需先将直流转换成交流。举例而言，这就意味着，能够在半个小时多一点点的时间内，完成三绕组变压器上两个低至1mHz的独立测量，而常规设计的旧仪器完成同样任务通常需要约两个小时的时间。

同样值得注意的是，具有合适介电频率响应测试设备的高压放大器，可将设备的测量能力扩展至在更高电压下的工频（50/60hz）电容和损耗角正切。测试设备能够测量的测试对象的最大电容取决于放大器的电流输出能力，而VAX020具有能供应50mA电流的放大器，则可对50Hz下电容高达80nF或60Hz下电容高达67nF的测试对象进行测试。除了测试变压器和套管之外，VAX200也是测量绝缘油的损耗正切角和/或电导的重要测试设置，测量这种绝缘油的国际标准通常推荐的测试电压为200到1000V/mm。

Megger公司VAX020放大器与本公司很受欢迎的高效IDAX300测试装置组合使用，是高压放大器和介电频率响应测试装置组合的最佳案例。这种组合的重量不超过10kg，将介电频率响应测试技术这一极有用的技术，可靠且低成本地拓展到之前很难甚至是不可能的应用中。

第五节　化学化工、生物、医学医药

一、中译英

　　人类个体一般在 20—35 岁时处于生命巅峰状态，随后进入衰老过程，表现为压力反应能力衰退，体内平衡失调增加以及患病风险提高。① 从生物学角度讲，衰老是指**老化的状态或过程**②，分为细胞衰老和机体衰老两个层次③。细胞衰老又叫"海弗里克极限"，由美国生物学家伦纳德·海弗里克于 1961 年发现，是指离体细胞表现出的分裂能力受限的现象；而机体衰老是指生物体的老化。衰老是已知的人类绝大多数疾病的**最大风险因素**④之一。一些学者把衰老当作一种疾病。由于基因对人类衰老的作用被发现，衰老正逐渐**被看做**⑤是可以治疗的遗传病。

　　人类寿命是由遗传因素和环境因素复杂的相互作用决定的。对**百岁老人**⑥以及模型动物的研究表明，一些单基因突变可大幅延长寿命。这些基因在哺乳动物

---

① 人类个体一般在 20—35 岁时处于生命巅峰状态，随后进入衰老过程，表现为压力反应能力衰退，体内平衡失调增加以及患病风险提高。Usually, human vitality reaches its peak at the age of 20—35 and then the aging process starts characterized by declined reaction capacity to pressure, increasing imbalance of internal environment and higher risk to suffer from diseases.

　　分析：本文有很多句子都用意译法翻译。本句译文意思相当于"人类生命力在 20—35 岁时达到巅峰，然后衰老过程开始，其特征为降低了的压力反应能力、越来越高的体内环境不平衡，以及更高的患病风险"。而且，译文前后两部分都是独立句子。后一句由于谓语太短，主语的定语 characterized by ... 被放到了谓语 starts 的后面。

　　在用词方面，原文三个名词＋动词的主谓词组"压力反应能力衰退，体内平衡失调增加以及患病风险提高"，被译成形容词（分词）＋名词形式。这也体现了中文动态，英文静态，中文多动词，英文多名词的特点。

② 老化的状态或过程：aged status or aging process。

　　注意：这里"老化状态"译成 aged status（已老化的状态），而"老化过程"则译成 aging process。

③ ……分为细胞衰老和机体衰老两个层次。... and comprises of cell aging and body aging.

　　分析：中文经常在讲完一件事情之后，在末尾加上补充性、概括性说明，这些补充性、概括性说明可以不译。本句的"两个层次"就是这样的例子。

④ 最大风险因素：the strongest risk factors。注意：这里的"最大"意思为"最强"，不宜译为 the largest。

⑤ 被看做：be viewed as。

⑥ 百岁老人：centenarian。

基因组中具有同源性。① 例如，这类基因既可以延长小鼠寿命，也与人类长寿有关。② 科学家已通过**基因操作**③，使模型动物寿命延长数倍。在未来，人类寿命也有望以基因技术手段得到显著延长。这是衰老机制的基因学说。

对于人类和哺乳动物，细胞衰老是由于端粒经过每次细胞分裂而不断变短所导致；当端粒变得太短时，细胞就会死亡。④ 端粒是染色体末端的 DNA 简单重复序列，人类和哺乳动物的端粒为 TTAGGG，重复 500～3000 次⑤。海弗里克指出，端粒是细胞的"分子钟"。

端粒酶是由 RNA 和蛋白质组成的复合体，属于反转录酶。⑥ 端粒酶的作用是**填补**⑦ DNA 复制缺陷，**维持端粒稳定**⑧，以增加细胞分裂次数，保持细胞活性。

统计学研究表明，端粒短的人群易患心脑血管病、糖尿病和癌症，寿命较短；

---

① 这些基因在哺乳动物基因组中具有同源性。In mammalian genomes, such genes are homologous.

分析：本句译文意思相当于"在哺乳动物中，这些基因是同源的"，其中"具有同源性"是典型的中式表达，其意思就是"同源的"。

② 这类基因既可以延长小鼠寿命，也与人类长寿有关 = 这类基因与小鼠和人类的长寿都有关。they are related to longevity of both mouse and human.

③ 基因操作 = 基因操纵：genetic manipulation。

④ 对于人类和哺乳动物，细胞衰老是由于端粒经过每次细胞分裂而不断变短所导致；当端粒变得太短时，细胞就会死亡。Aging of human and mammals is caused by continuously shortening of telomeres every time the cells divide. Cells die when the telomeres are short enough.

分析：本句译文意思相当于"人类和哺乳动物的衰老是由于端粒在每次细胞分裂时都持续变短所致。当端粒太短时，细胞就会死亡"。经过这样的转化后，译文简洁了很多，而且更符合英文行文习惯。

⑤ 人类和哺乳动物的端粒为 TTAGGG，重复 500–3000 次。For human and mammals, telomere is TTAGGG repeating for 500–3000 times.

分析：本句的意思，实际上是说，人类和哺乳动物的端粒为 TTAGGG，这个端粒被重复了 500–3000次。

⑥ 端粒酶是由 RNA 和蛋白质组成的复合体，属于反转录酶。Telomerase is a reverse transcriptase and complex of RNA and protein.

分析：本句译文实际上是将"是……的复合体，属于反转录酶"理解成"是反转录酶，也是……的复合体"。

⑦ 填补 = 补偿：make up for。

⑧ 维持……稳定 = 使稳定：stabilize ...

而端粒长的人群更健康，寿命更长。① 另有研究发现，生物抗氧化物质**对**端粒和端粒酶**具有保护作用**②，有利于健康长寿。这是衰老机制的端粒学说。

Usually, human vitality reaches its peak at the age of 20—35 and then the aging process starts characterized by declined reaction capacity to pressure, increasing imbalance of internal environment and higher risk to suffer from diseases. In biology, aging refers to aged status or aging process and comprises of cell aging and body aging. The cell aging is also known as "Hayflick limit", which was found by Leonard Hayflick, an American biologist, in 1961. It refers to division ability of isolated cells is limited. The body aging refers to the aging of a living body. Aging is one of the strongest risk factors for most known human diseases. Some researchers even consider aging itself a disease. Since effect of gene on aging has been discovered, aging is gradually viewed as a curable genetic disease.

Human life is determined by the complex interaction between genetic factors and environmental factors. Researches on centenarians and model animals prove that some single gene mutations may greatly prolong life. In mammalian genomes, such genes are homologous. For example, they are related to longevity of both mouse and human. Scientists have prolonged the lives of model animals by many times by genetic manipulation. In the future, human life may be considerably prolonged via gene technology. This is the gene theory of aging mechanism.

Aging of human and mammals is caused by continuously shortening of telomeres every time the cells divide. Cells die when the telomeres are short enough. The telomere is a simple repetitive sequence at the end of the chromosome. For human and

---

① 统计学研究表明，端粒短的人群易患心脑血管病、糖尿病和癌症，寿命较短；而端粒长的人群更健康，寿命更长。Statistics researches show that people with short telomeres are vulnerable to cardia-cere-brovascular diseases, diabetes and cancers and short-lived while people with long telomeres enjoy better health and longer lives.

分析：译文第一句将"易患……和……"与"寿命较短"译成 be vulnerable to ... and ... and short-lived，也就是说，将"寿命较短"译成 be short-lived，不译成 with short life，这样就可以与前面的 be vulnerable to 共用 be 动词，显得更简洁。第二句译文则相当于"更健康，寿命更长"理解成"享有更好的健康和更长的寿命"。

② 对……具有保护作用＝能保护……：can protect。

mammals, telomere is TTAGGG repeating for 500—3000 times. Hayflick points out that the telomere functions as a "molecular clock" of a cell.

Telomerase is a reverse transcriptase and complex of RNA and protein. It makes up for DNA replication defectives and stabilizes the telomeres so as to increase division times and maintain cell viability.

Statistics researches show that people with short telomeres are vulnerable to cardia-cerebrovascular diseases, diabetes and cancers and short-lived while people with long telomeres enjoy better health and longer lives. Other researches find that biological antioxidants can protect the telomeres and telomerase, in favor of health and longevity. This is the telomeres theory of aging mechanism.

二、英译中

Lubrication

Proper lubrication is vital to compressor operation and requires special attention in **package**[①] design. Two independent systems lubricate a compressor: the **frame oil system**[②] and the **force feed system**[③]. The frame oil system is a pressurized circulating system that supplies oil to the crankshaft, connecting rods, and crossheads. The force feed system is a high-pressure injection system that supplies small quantities of oil to the piston rod packings and piston rings.

---

① package：*n.* 成撬。
说明：成撬就是指把各个不同的配件组装成一个成套机器。成撬商就是负责把各个不同厂家的配件组装成一个成套机器的厂家。
② frame oil system：机身润滑油系统。frame：*n.* 机身。
③ force feed system = force feed lubrication system：压力润滑系统。
分析：force feed system 在网络上难以找到其中文解释，但可以找到很多 force feed lubrication system，这里根据上下文语境推测，force feed system 就是 force feed lubrication system（压力润滑系统）。根据本段后文可知，这种推测是合理的。

In a compressor, lubrication:①

1. Reduces friction—Decreases energy consumption and heat generation.

2. Reduces **wear**②—Increases equip-ment life and decreases maintenance costs.

3. Removes heat from the system—Cools moving parts and maintains **working clearances**③.

4. Prevents corrosion—Generally provided by **additives** rather than the **base lubricant**.④

5. Seals and reduces contaminant **buildup**⑤—Improves gas seal on piston and packing rings, and flushes away contaminants from moving parts.

6. Dampens shock—Reduces vibration and noise and increases component life.

---

① In a compressor, lubrication: …… 在压缩机中，润滑油的作用包括……

分析：本句其实不是个完整句，只是分点式句子的共同部分，后面还跟着6个小点。而且这6个小点都是以动词+名词作为小标题，后面用破折号说明具体内容。更重要的是，lubrication 与6个小点的动词+名词刚好成为一个完整的主谓宾结构。因此，在翻译时，宜根据中文行文习惯，其译成"润滑油的作用包括……"这是根据原文意思，灵活组织译文，使译文符合目标语行文习惯的典型例子。

另外，本部分内容的6个小点，原文都是动词+名词并加解释的格式。因此，译文也宜在准确的基础上，统一采用动词+名词并在破折号后加解释的格式。如果不采用这种格式，就会感觉译文不协调。这应该也算是技术类翻译的某种程度上的"雅"。

② wear：*n*. 磨损；损耗；耐久性。

③ Cools moving parts and maintains working clearances. 冷却活动部件，维持机器的工作间隙。

分析：本句隐含着一个意思，即让活动部件冷却下来，因为部件受热膨胀后，让机器的工作间隙变小，工作间隙自然就无法维持。

④ Prevents corrosion—Generally provided by additives rather than the base lubricant. 防止腐蚀——腐蚀一般是由添加剂，而非润滑油基础油造成。

分析：本句跟其他5个小点一样，也是小标题加破折号后解释的格式。但本句与其他5个小点不同的是，本句破折号后的内容是 corrosion 的后置定语。句子的意思是润滑油可防止由添加剂（而非基础油）造成的腐蚀。

additive：*n*. 添加剂。

base lubricant：润滑油基础油。

背景知识：润滑油基础油主要分矿物基础油、合成基础油以及植物油基础油三类。矿物基础油应用广泛，用量很大（约90%以上），但有些应用场合则必须使用合成基础油和植物油基础油调配的产品，酯类油作为滑油高端使用。

⑤ buildup：*n*. 累积；积累。

Many types of oils exist, some petroleum based, others synthetic.① Each oil exhibits different characteristics that suit it for a specific application.

Lubricant Terminology②

VISCOSITY—Measures fluid resis-tance to flow.③ It decreases with increasing temperature. In this document, viscosity is expressed in centistokes (cSt)④. Proper viscosity is the most important aspect of compressor lubri-cation. FIGURE 6-4 illustrates viscosity differences between base stock types.

Viscosity can increase with **oxidation** or contamination by a liquid of higher viscosity or decrease with contamination by hydrocarbon gas condensate or other liquid of lower viscosity.⑤ Oil **degradation**⑥ increases viscosity, **unless it is**⑦ multi-viscosity

---

① Many types of oils exist, some petroleum based, others synthetic. 润滑油种类繁多，其中有些为石油基润滑油，其他的则为合成润滑油。

分析：本句实际上相当于：Many types of oils exist, some are petroleum based, and others are synthetic.

② Terminology：n 术语。

③ VISCOSITY—Measures fluid resistance to flow. 黏度——用于衡量液体流动阻力的参数。

分析：本句看起来似乎是一个标题后面跟着一个不完整句子，但实际上，这是一个完整句子，其主语就是标题 VISCOSITY，破折号后的内容则是谓语和宾语。像这样的句子结构，在翻译时也应根据情况，尽可能让译文读起来像个句子。另外，VISCOSITY 所有字母都大写，起到了突出显示的作用，但中文无大小写之分，作为补偿，可以将 VISCOSITY 译文加粗。

④ viscosity is expressed in centistokes (cSt) 黏度单位为厘泊（cSt）。

分析：本句不宜字面译成"黏度以厘泊（cSt）表达"，而是应根据语境译成"黏度单位为厘泊（cSt）"，才会符合中文行文习惯。另外，厘泊的英文缩写 cSt 中间的 S 大小，另外两个字母小写，这才是规范的写法。因为泊是黏度的基本单位，cSt 里的 c 是"厘"（百分之一）的意思，不能大写。

⑤ Viscosity can increase with oxidation or contamination by a liquid of higher viscosity or decrease with contamination by hydrocarbon gas condensate or other liquid of lower viscosity. 润滑油被氧化或受高黏度液体污染后会导致黏度升高，受烃类气体冷凝物或其他低黏度液体的污染则会降低黏度。

分析：本句结构为 Viscosity can increase (with ...) or decrease (with ...)。其中第一个 with 后面有两个宾语 oxidation or contamination，而且后面一个宾语有个 by 引出一个施动者 a liquid of higher viscosity。第二个 with 后只有一个宾语 contamination，但 contamination 后面的 by 引出的施动者有两个，即 hydrocarbon gas condensate 和 other liquid of lower viscosity。

Oxidation：n. 氧化；氧化反应。这里指受到氧化。

⑥ degradation：n. 降解。

⑦ unless it is：除非其为；但……例外。

oil[①] (such as SAE 10W40). In multi-viscosity oils, the **viscosity improvers**[②] degrade, not the base oil itself.

VISCOSITY INDEX—Indicates the **magnitude**[③] of viscosity change with respect to temperature. The higher the viscosity index, the less viscosity decreases as temperature increases.

POUR POINT—Specifies the lowest temperature at which oil flows. It is important in cold weather applications and in cylinder and packing lubrication with cold suction temperatures.

FLASH POINT—Specifies the lowest temperature at which oil vaporizes to create a combustible mixture in air. If **exposed to**[④] flame or high temperature, the mixture **flashes into flame**[⑤] and then extinguishes itself. This is important in high temperature applications where oil may mix with air.

润滑

适当的润滑对于压缩机作业来说至关重要，并且在成橇设计中需要特别关注。压缩机的润滑需要使用两个独立的系统，即机身润滑油系统与压力润滑系统。其中，机身润滑油系统是加压循环系统，可为曲轴、连杆以及十字头提供润滑油。压力润滑系统是高压注油系统，可向活塞杆填料以及活塞环提供少量的润滑油。

在压缩机中，润滑油的作用包括：

1. 减少摩擦——降低能量消耗与热量的产生。

2. 减少磨损——延长设备寿命，降低维护成本。

3. 去除系统内部热量——冷却活动部件，维持机器的工作间隙。

4. 防止腐蚀——腐蚀一般是由添加剂，而非润滑油基础油造成。

5. 密封并减少污染累积——改善活塞与填料环的气封，并冲洗掉活动部件中的污染物。

---

① multi-viscosity oil：多黏度润滑油。
背景知识：多黏度润滑油内含多种殊添加剂，低温时易流动、不凝结，高温时保持黏稠度、不分解。

② viscosity improver：黏度改进剂。

③ magnitude：*n.* 大小；量级。

④ be exposed to：接触；暴露于。

⑤ flash into flame：闪燃。

6. 缓解冲击——减少振动与噪声，延长零部件寿命。

润滑油种类繁多，其中有些为石油基润滑油，其他则为合成润滑油。每种润滑油都有其不同的特性，适用于特定的用途。

润滑油术语

黏度——用于衡量液体流动阻力的参数。黏度随着温度升高而降低。在本文中，黏度单位为厘泊（cSt）。适当的黏度是压缩机润滑作用所需的最重要特性。图 6-4 展示了不同基础油类型的粘度差异。

润滑油被氧化或受高黏度液体污染后会导致黏度升高，受烃类气体冷凝物或其他低黏度液体的污染则会降低黏度。润滑油降解后黏度会升高，但多黏度润滑油（如 SAE 10W40）例外。在多黏度润滑油中，发生降解的是黏度改进剂，而非润滑油基础油本身。

黏度指数——表示黏度随温度变化的程度。黏度指数越高，表示黏度受温度的影响越小。

倾点——表示润滑油能够流动的最低温度。当在低温条件下使用，或者在对要求低吸入温度的气缸与填料进行润滑时，润滑油的倾点是很重要的性质。

闪点——表示润滑油蒸发并在空气中产生可燃混合物的最低温度。如果该混合物接触火焰或高温，则会闪燃，而后自我熄灭。对于高温条件下使用并且润滑油可能与空气混合的情况下，闪点十分重要的性质。

第六节　石油地质采矿

一、中译英

喷射法[①]**下**[②]表层**导管**[③]是为适应深水海床土质疏松的特点和避免浅层地质灾

---

① 喷射法 = 深水喷射钻井法：deepwater jetting drilling。
　说明：这里的"喷射法"，根据上下文语境，宜理解成"深水喷射钻井法"。
② 下：run；lower；set。
　说明：钻井中的"下"，是指将套管、导管等从地面放到井内。
③ 导管：conduit。
　说明：石油钻井导管是一段大直径的短套管，用在沼泽地井场上和其他特定环境下；其主要作用是保持井口敞开，防止钻井液冲出表面地层，可将上溢的钻井液传输到地面。

害而发展起来的深水钻井技术。该技术的基本工艺为：作业时将导管和**钻具**①**组合**②为一体，通常由**牙轮钻头**③水眼喷射出的水流冲刷土体形成钻孔，导管和钻具的组合体利用自重**下入**④至设计深度，然后解脱导管和钻具，完成**二开井眼钻进**⑤后**上提**⑥钻具并将钻具送至钻井船。在喷射钻进过程中，钻头底部伸出**导管鞋**⑦的长度称为**钻头伸出量**⑧，它始终保持不变，即钻头与导管在竖直方向上成为一体，同步下入地层。

在喷射钻井中，**钻井液**⑨通过钻头**喷嘴**⑩喷射而产生水射流，以实现**井下**⑪的**破岩和清岩**⑫。钻井液经钻头喷嘴流出后，不仅被**井筒**⑬中的液体淹没，而且还受到**井壁**⑭和井底的限制以及反喷射流的干扰，所以从钻头喷嘴流出的射流是淹没非自由射流。此外，由于钻头上的喷嘴均为圆形，所以这种射流又属于圆形射流，具有轴对称性。

为建立合理钻头伸出量计算模型，首先分析不同钻头伸出量产生的喷射钻进效果。当钻头伸出量过小时（钻头喷嘴处于导管鞋上部且轴向距离大），射流束将部分或全部**作用**⑮在导管内壁，不能充分破碎土体和清洁井底，导致成孔和钻进效率低下；当钻头伸出量过大时（钻头喷嘴与导管鞋轴向距离小甚至处于导管鞋下部），射流束将会过度冲刷破坏钻头底部土体，使破碎的泥屑从导管外侧**返**

---

① 钻具：drilling tools。
② 组合：integrate。
③ 牙轮钻头：roller bit。钻头：bit。
④ 下入：set；lower；run。
⑤ 二开井眼钻进 = 二开井井眼钻进：the second spudding of well hole。spudding：n. 开钻。
⑥ （钻具等的）上提：lift。
⑦ 导管鞋：conduit shoes。
⑧ 钻头伸出量：bit stick-out。
说明：这个词的译法可在百度里查"钻头伸出量 bit"找到。
⑨ 钻井液：drilling fluid。
说明：钻井液，是钻井过程中以其多种功能满足钻井工作需要的各种循环流体总称。
⑩ 喷嘴：nozzle。
⑪ 井下（井内）：downhole。
⑫ 破岩和清岩 = 岩石破碎和清除：rock breaking and clearing。
⑬ 井筒：borehole。
⑭ 井壁：sidewall。
⑮ 作用 = 施加：apply。

出①，从而导致导管喷射到位后周围土体回填困难，使导管无法及时获得足够的地层承载力②，影响水下井口③稳定性。基于上述分析，假设当射流束上边缘与导管下边缘内侧相切时钻头伸出量最优，即射流束上边缘 QG 与导管内壁 AG 相切于 G 点，此时既能保证导管喷射下入速度，又能满足安全作业要求。④ 本文根据这一理论假设建立了合理钻头伸出量的计算模型。

The running of surface conduit for deepwater jetting drilling is a deepwater drilling technology developed for adapting to the characteristics of loose deepwater seabed soil and preventing shallow geological disasters. The basic process of this technology: the conduit and drilling tools are integrated during the operation, water flow generally sprayed by water nozzle tips of roller bit flushes the soil body to form a drill hole, and the combination of conduit and drilling tools are set to the design depth by their dead weight, and then released; after finishing the second spudding of well hole, drilling tools are lifted and sent to a drilling ship. During jetting drilling, the length of bit bottom extending the conduit shoes is called bit stick-out, it is kept constant, that is, the bit and the conduit integrated into a whole in vertical direction, and run into the formation simultaneously.

During jetting drilling, drilling fluid is sprayed by bit nozzle and produces water jet for downhole rock breaking and clearing. The drilling fluid flown out by the bit is not only submerged by the fluid from the borehole, but also restricted by the sidewall

---

① 返出：return。

② 承载力：bearing force。

③ 井口：wellhead。

④ 基于上述分析，假设当射流束上边缘与导管下边缘内侧相切时钻头伸出量最优，即射流束上边缘 QG 与导管内壁 AG 相切于 G 点，此时既能保证导管喷射下入速度，又能满足安全作业要求。Based on the above analysis, it is assumed that the bit stick-out is the best when the upper edge of jet beam is internally tangent to the lower edge of conduit, that is, the upper edge QG of jet beam is tangent to inner wall AG of conduit at point G, which can not only guarantee jetting running speed of conduit but also satisfy safety operation requirements.

分析：本句主干是"假设钻头伸出量最优"，后面的"即射流束上边缘 QG 与……作业要求"是对主句的补充说明。翻译时，宜先译出主句，再译后面的补充说明，其中"此时既能……又能……"可以译成 which 引导的非限制性定语从句。

and downhole and interfered by reflection jet, thus the jet from the bit nozzle is a submerged non-free jet. Moreover, all bit nozzles are round, so this jet is also a circular jet, and axially symmetric.

To build a reasonable bit stick-out calculation model, jetting drilling effects generated at different bit stick-outs are analyzed first. When bit stick-out is too small (bit nozzles locate at the upper part of conduit shoe and have large axial distance with the conduit shoe), jet beam will be partially or wholly applied to the internal wall of conduit, and cannot fully break soil body and clean the downhole, resulting in low pore-forming and drilling efficiency; when the bit stick-out is too large (bit nozzles locate at the lower part of conduit shoe and have large axial distance with the conduit shoe), the jet beam will excessively flushes and breaks the soil body of bit bottom, and the broken mud-sized grain returns from outside of the conduit, so the refilling of surrounding soil body is difficult when the conduit jetting is in place, and the conduit has no sufficient formation bearing force, impacting the stability of subsea wellhead. Based on the above analysis, it is assumed that the bit stick-out is the best when the upper edge of jet beam is internally tangent to the lower edge of conduit, that is, the upper edge QG of jet beam is tangent to inner wall AG of conduit at point G, which can not only guarantee jetting running speed of conduit but also satisfy safety operation requirements. With this theoretical assumption, the paper establishes a model for calculating reasonable bit stick-out.

二、英译中 [①]

Overturning and Sliding

The maximum allowable static **coefficient of friction** to be used in **overturning** or inadvertent **rig sliding** calculations of drilling structures supported by soil, concrete, or **wood matting foundations** shall be limited to 0.15, and to 0.12 for those supported by steel foundations, except as follows: **alternative values** for the maximum design coefficient of friction may be used **provided** such values have been **validated thru** testing and are consistent with rig operating procedures (e.g. an offshore **skiddable**

---

① 本部分内容难度较大，大家可选择性学习。

rig design **incorporating** a coefficient of friction consistent with **ungreased** surfaces would require that the owner/operator maintain and inspect the beams to ensure that they are not inadvertently **greased**).①

For all stability and sliding calculations, **dead weights**② providing resistance to overturning or sliding shall be limited to a maximum of 90 % of their expected minimum weight. The calculation of minimum weight shall assume the removal of all **optional** structures and equipment, and **fluid tanks** shall be **considered** empty, **unless**

---

① The maximum allowable static **coefficient** of friction to be used in **overturning** or inadvertent **rig sliding** calculations of drilling structures supported by soil, concrete, or **wood matting foundations** shall be limited to 0.15, and to 0.12 for those supported by steel foundations, except as follows: **alternative values** for the maximum design coefficient of friction may be used **provided** such values have been **validated thru** testing and are consistent with rig operating procedures (e.g. an offshore **skiddable** rig design **incorporating** a coefficient of friction consistent with **ungreased** surfaces would require that the owner/operator maintain and inspect the beams to ensure that they are not inadvertently **greased**). 在土壤、混凝土或木垫基础所支承钻井结构的钻机倾覆或意外滑动的计算中，所使用的最大许用静态摩擦系数的极限为 0.15，对于钢基础所支承的钻井结构，该极限为 0.12，但以下情况例外：如果其值通过试验得到确认，并与钻机作业程序（例如，海上滑橇式钻机设计，其摩擦系数对应的表面未润滑，要求业主/经营者维护和检验横梁，以确保其未被意外润滑）相一致，则最大设计摩擦系数可以使用替代值。

分析：本句整段就是一个句子。其结构如下：

a. except as follows 及其后面部分为全句的条件状语，except 前面部分为主句。

b. 主句结构为 The maximum allowable static coefficient (of friction to be used in ... or wood matting foundations) shall be limited to 0.15, and to 0.12 (for those ...)。其中 in 后的宾语相当于 overturning calculation and inadvertent rig sliding calculation，但为避免重复，省略了前面的 calculation，但后面一个改为复数形式。of drilling structures supported by soil, concrete, or wood matting foundations 是 calculations 的后置定语。

c. except 所引导的条件状语主干为 alternative values may be used。

d. 条件状语从句后面 provided 引导的是让步状语从句，其主干为 such values have been validated and are consistent with rig operating procedures。

e. 最后面括号部分也是个复杂句，其主干为 an offshore skiddable rig design would require that ... 其中 that 从句后面还有一个 that 从句，后面一个 that 从句是目的状语从句。

coefficient of friction：摩擦系数。overturning：*n.* 倾覆。rig：*n.* 钻机。sliding：*n.* 滑动。

wood matting foundations：木垫基础。foundation：*n.* （建筑物；结构物）基础。

alternative value：替代值。provided (that)：*conj.* 如果；前提是。

validate：*vt.* 确认；验证；使生效。thru = through。

skiddable：*adj.* 可滑动的；滑橇式的。incorporating = having 或 containing。

ungreased：*adj.* 未润滑过的；未上润滑油的。greased：*adj.* 润滑过的；上过润滑油的。

② dead weight：自重；静负载；固定负载。

otherwise specified in the rig instructions for **storm preparations** or **rig erection**.[①] For drilling structures **subject to**[②] vertical **heave**[③], the stabilizing weights shall be further reduced by the [④]**magnitude** of the negative heave acceleration.

Freestanding structures on land shall have a minimum **factor-of-safety** against overturning of 1.25, **calculated as the ratio of** the minimum stabilizing **moment** of the dead weight of the structure, taken about a **tipping line, divided by** the overturning moment of the sum of any overhanging vertical **live loads** plus environmental loads,

---

① The calculation of minimum weight shall assume the removal of all **optional** structures and equipment, and **fluid tanks** shall be **considered** empty, **unless otherwise specified in** the rig instructions for **storm preparations** or **rig erection**. 计算最小重量时，应假定除去所有非必须的结构和设备，并假定油箱是空的，除非钻机说明书中对预防风暴或钻机安装另有规定。

the removal of all optional structures and equipment：除去所有非必须的结构和设备。optional：*adj.* 非必需的。

consider：*vt.* 假定。fluid tank：油箱。

storm preparations：预防风暴。unless otherwise specified in …：除非在……中另有规定。

② subject to：会产生……（不译成"承受"）。

③ heave：*n.* 升沉；起伏。

④ magnitude：*n.* 大小；量级。

including wind or earthquake, taken about the same tipping line or axis.[①] The designer shall consider suitable tipping lines so as to determine the minimum factor of safety and shall consider the possibility of overturning loads from any possible **direction of application**[②]. Determination of the location of a tipping line **shall be such that**[③] the tipping line shall lie along the **centroid**[④] of the nominal vertical ground support **reactions**[⑤] for the case considered; the distribution of ground support reactions shall be limited to comply with design allowable ground bearing pressures for the

---

① Freestanding structures on land shall have a minimum **factor-of-safety** against overturning of 1.25, **calculated** as the **ratio of** the minimum stabilizing **moment** of the dead weight of the structure, taken about a **tipping line**, **divided by** the overturning moment of the sum of any overhanging vertical live loads plus environmental loads, including wind or earthquake, taken about the same tipping line or axis. 陆上独立式结构的最低倾覆安全系数为1.25，即倾覆线附近结构自重的最小稳定力矩，除以同一倾覆线或倾覆轴附近任何悬挂的垂直活载加上环境载荷（包括风或地震载荷）之和的倾覆力矩的比值。

分析：本句结构分析如下：

a. 主句为第一小句，后面的 (which is) calculated as ... 是 a minimum factor-of-safety against overturning 的定语。

b. calculated as 后的所有内容相当于一个名词结构，其结构为 the ratio of A divided by B (of the sum of C plus D)，其意思为 A 除以 B 和 C 之和的比值。

c. the ratio of A divided by the sum of B plus C 结构中，A 为 the minimum stabilizing moment of the dead weight of the structure，后面的 (which is) taken about a tipping line 是 the minimum stabilizing moment 的后置定语。B 为 the overturning moment，后面的 of the sum of C plus D（即 any overhanging vertical live loads）为 B 的后置定语。

d. 最后面的 including wind or earthquake 是 environment loads 的举例，而 taken about the same tipping line or axis 则是 wind or earthquake (loads) 的后置定语。

freestanding structures：独立结构。factor-of-safety：安全系数。

calculated as：计算成；计为。这里理解为"即"。

moment：*n.* 力矩。tipping line：倾覆线。

the ratio of A divided by B (of the sum of C plus D)：A 除以 C 和 D 之和的 B 的比值。

live load：活载荷。

② direction of application：受力方向；施力方向。application：*n.*（力的）施加。

③ shall be such that：应为这样，以使得……

④ centroid：*n.* 质心；形心；几何中心。

⑤ reaction：*n.* 反作用力；反作用。

structures under consideration.① The manufacturer shall **include**② foundation loading diagrams and the required safe ground bearing pressure allowable for erection and operating conditions **in** the rig manual. Freestanding land drilling structures shall have a minimum factor of safety against inadvertent sliding of 1.25, calculated as the ratio of the minimum sliding resistance at the design maximum allowable static coefficient of friction, divided by the total **applied**③ **shear loads**④ due to environmental loads.

倾覆和滑动

在土壤、混凝土或木垫基础所支承钻井结构的钻机倾覆或意外滑动的计算中，所使用的最大许用静态摩擦系数的极限应为0.15，对于钢基础所支承的钻井结构，该极限应为0.12，但以下情况例外：如果其值通过试验得到确认，并与钻机作业程序（例如，海上滑橇式钻机设计，其摩擦系数对应的表面未润滑，要求业主/经营者维护和检验横梁，以确保其未被意外润滑）相一致，则最大设计摩擦系数可以使用替代值。

对于所有稳定性和滑动计算，产生倾覆或滑动阻力的自重最大值应限定为其预期最小重量的90%。计算最小重量时，应假定除去所有非必须的结构和设备，并假定油箱是空的，除非钻机说明书中对预防风暴或钻机安装另有规定。对于会产生垂直升沉的钻井结构，还应按照负升沉加速度的大小，将稳定重量进一步降低。

陆上独立式结构的最低倾覆安全系数为1.25，即倾覆线附近结构自重的最小

---

① the distribution of ground support reactions shall be limited to comply with design **allowable ground bearing pressures** for the structures **under consideration**. 应对地面支承反作用力的分布进行限制，使其符合所考虑结构的设计许用地面承受压力。

分析：本句主干为 the distribution of ground support reactions shall be limited，后面的 to comply with ... under consideration 是主句的目的状语。

allowable ground bearing pressures：许用地面承受压力。under consideration：所考虑的；考虑中的。

② include A in B：将 A 放入 B 中；B 中包括 A。

说明：the manufacturer shall include ... in the rig manual 的字面意思是"制造商应将……包括在钻机手册内"，但应意译成"制造商的钻机手册应包括……"这种英文表达法是一种很地道的句式，在中英翻译时也可以借鉴使用。

③ applied：*adj.* 所施加的（力）；所作用的（力）。

④ shear load：剪切载荷。

稳定力矩，除以同一倾覆线或倾覆轴附近任何悬挂的垂直活载加上环境载荷（包括风或地震载荷）之和的倾覆力矩的比值。设计人员应选用合适的倾覆线，以便确定最低安全系数，并应考虑因任何可能受力方向的可能倾覆载荷。倾覆线位置应这样确定，倾覆线应经过所考虑情况地面公称垂直支承反作用力的质心；应对地面支承反作用力的分布进行限制，使其符合所考虑结构的设计许用地面承受压力。制造商的钻机手册内容应包括基础载荷图，以及建造和作业条件下要求的许用安全地面承受压力。陆上独立式钻井结构的最低意外滑动安全系数为1.25，即设计最大许用静摩擦系数下的最小滑动阻力除以由环境载荷施加的总剪切载荷的比值。

科技翻译教程

尊敬的老师:

您好!

为了方便您更好地使用本教材,获得最佳教学效果,我们特向使用该书作为教材的教师赠送本教材配套参考资料。如有需要,请完整填写"教师联系表"并加盖所在单位系(院)公章,免费向出版社索取。

北京大学出版社

教 师 联 系 表

教材名称		科技翻译教程		
姓名:	性别:		职务:	职称:
E-mail:		联系电话:		邮政编码:
供职学校:		所在院系:		(章)
学校地址:				
教学科目与年级:		班级人数:		
通信地址:				

填写完毕后,请将此表邮寄给我们,我们将为您免费寄送本教材配套资料,谢谢!

北京市海淀区成府路 205 号
北京大学出版社外语编辑部　刘文静
邮政编码: 100871
电子邮箱: zpup@pup.cn

邮 购 部 电 话: 010-62534449
市场营销部电话: 010-62750672
外语编辑部电话: 010-62754382